ENVIRONMENTAL CHANGE AND HUMAN DEVELOPMENT

ENVIRONMENTAL CHANGE AND HUMAN DEVELOPMENT

Controlling Nature?

C.J. BARROW

School of Social Sciences and International Development,
University of Wales Swansea, UK

A MEMBER OF THE HODDER HEADLINE GROUP

First published in Great Britain in 2003 by
Hodder Arnold, an imprint of Hodder Education and
a member of the Hodder Headline Group,
an Hachette Livre UK Company
338 Euston Road, London NW1 3BH

www.hoddereducation.com

The advice and information in this book are believed to be true and
accurate at the date of going to press, but neither the author nor the publisher
can accept any legal responsibility or liability for any errors or omissions.

British Library Cataloguing in Publication Data
A catalogue record for this book is available from the British Library

Library of Congress Cataloging-in-Publication Data
A catalog record for this book is available from the Library of Congress

ISBN 978 0 340 76404 6

Typeset in 10/12pt Sabon by Phoenix Photosetting, Chatham, Kent

What do you think about this book? Or any other Arnold title?
Please visit our website: www.hoddereducation.com

Contents

Preface

This book focuses on environmental change and human fortunes. While there is a large and rapidly expanding literature dealing with how people affect the environment, less attention has been given in recent years to how the environment shapes human development. In an ever more crowded world there is a need for anticipatory environmental management, and a crucial input to this is consideration of the interaction between environment and humans.

Nowadays, planners, administrators and the people they serve are often oblivious to the fact that the environment is a key control on the quality of life. In richer nations, people commonly feel that modern technology gives them 'the edge' over nature to a degree never enjoyed before; and in poorer countries established ways of coping with natural challenges have often broken down, and new technological solutions are beyond easy access. City folk overlook the fact that each modern high-rise building, towering above a hectare plot (or less), has enormous 'roots' stretching out to oil-fields, many square kilometres of croplands, ocean fisheries, cattle ranches, rivers and reservoirs, and so on – all dependent upon and vulnerable to nature.

The environment is not as stable, benign or controllable as many like to think. The world population is vastly larger than it ever has been and is still growing, and humans increasingly upset nature through pollution and other activities. While modern communications may help environmental managers, rapid travel also increases the dispersal of diseases and pests. Technological advance and social development are not all beneficial; some innovations have the effect of making people more vulnerable to disruption by natural disaster, and citizens are often less able to cope with and adapt to changed conditions than people were in the past.

In the modern world there is, in practice, relatively limited interest from governments and voters in anticipatory studies, contingency planning and adaptive strategies. Over the past few decades there has been some shift towards concern for environmental issues, stimulated by fears about global warming, stratospheric ozone depletion, pesticide pollution and, by the late

1990s, El Niño events and the associated forest fires in Southeast Asia and Latin America, and flooding in the UK; however, the focus has been mainly on *how humans are affecting the environment*, and much less on *how the environment can affect humans*. This book, I hope, will help prompt consideration of the latter, and development that includes vulnerability reduction, and improvement of flexibility and adaptability.

There are three sections in this book: the first (Chapters 1 to 3) deals with debates about environment–human interaction, the attitudes people have towards the environment, and past environmental impacts; the second section (Chapters 4 to 5) focuses on recent and ongoing changes, asking whether we are more or less vulnerable than in the past; section three (Chapters 6 to 8) reviews likely future challenges and explores how to ensure a more secure, as well as materially improved human future.

Inevitably there is a tendency to report past catastrophes and to flag threats, but I am also keen to identify opportunities and strategies to improve the chances of achieving better human welfare.

Chris Barrow
University of Wales Swansea
Swansea, UK
May 2002

Acknowledgements

I wish to thank the editorial team at Arnold, in particular Liz Gooster and those who kindly read my draft. I am also most grateful to staff of the Inter-Library Loans Office, University of Wales Swansea.

1

Introduction

It is possible to think of environmental history as being about two relatively fixed and largely independent elements – the humans and the natural – working and interacting together. ... But what happens when this symbiotic relationship breaks down or is fundamentally altered because the environment itself changes or because people migrate to new, environmentally very different lands? (Arnold, 1996: 98)

Chapter summary

This chapter explores the environment–human relationship, focusing on human perceptions of nature. Particular attention is paid to attempts to relate environment to development and an attempt is made to review relevant concepts, notably: ideas of Eden, Utopia and Arcadia; environmental determinism and possibilism; catastrophism and uniformitarianism; and the Gaia Hypothesis.

With newspapers, television and other media regularly featuring matters of environmental concern it may seem redundant for me to call for more awareness of environmental issues. However, I suggest that too much attention is presently directed toward human impacts upon the environment, especially global warming, and not enough toward how nature affects people. Issues like soil degradation, natural climatic variations, volcanic activity, and many other environmental threats and opportunities deserve far more consideration. Many of the world's governments pursue strategies and build infrastructure making the assumption that nature is essentially stable and benign. The reality is that the Earth has never had a constant, favourable environment – change has been the norm (Mannion, 1999: 1).

Decision-makers tend to respond to problems when they become clearly manifest, and neglect anticipatory and contingency planning. The disastrous floods in the UK during 2000 were no surprise to hydrologists and many other environmental specialists, yet the media, the public and the government have tended to see them as unusual, or a consequence of 'global warming', and the impact has been far more costly than it should have been. Modern humans are frequently much less adaptive than their ancestors and this is a serious disadvantage. Many communities today are moving toward or are already locked into livelihood strategies, infrastructure and food dependency, which are sensible and secure only if the environment is reasonably stable and unchanging. History, palaeoecology, archaeology and numerous other disciplines warn clearly that the environment is often unstable and subject to change, and that it will fluctuate markedly even without humans upsetting natural systems, although they are doing that also. People may seem to be protected by technological progress, but the reality is that vulnerability to environmental change has increased, because of one or more of the following factors:

- There are larger human populations than ever before.
- People depend upon more complex, easily disrupted and difficult to repair infrastructure, services and administration.
- In the past people have had the option of migration if conditions became unfavourable, with a fair chance that there would be somewhere to allow them a new start; today most lands are densely settled so that is unlikely.
- Rapid modern transport can disperse diseases and pests more effectively than ever before. In the past poor communications helped put communities in protective quarantine.
- In many nations people can no longer draw on help from extended families and local communities in times of need.
- Individuals in richer countries have less ability to fend for themselves, because specialists are needed to install and repair infrastructure and provide services they depend upon.
- There are often weaker social and moral supports.
- Resistance to infection may have been reduced through better hygiene and medicine, and possibly pollution.
- People are often less used to hardship, and have few practical survival skills.

Humans have come to dominate all other multicellular organisms by virtue of being capable of forward planning, and through adaptability and resilience. Nowadays, these qualities are in need of reinforcement. Also, decision-makers and environmental managers need to know more about how individuals, local groups and governments react to environmental challenges and opportunities. Misery may not only be directly caused by environmental change, human responses can add serious indirect effects: for example, a natural disaster might prompt warfare.

Humans are increasing their capacity to modify the genetic make-up of themselves and other organisms, potentially a way of increasing adaptability through developing new crops more effectively and rapidly, improving countermeasures against disease, and even developing new pollution control and soil improvement measures (these developments should help counter increasing vulnerability; see Chapter 4). Technology, including biotechnology, increasingly offers ways of fighting environmental threats and of exploiting opportunities, but this is only likely to be effective if people sufficiently grasp the need to develop it in the right ways and are willing to pay to do that. Since the 1950s the sciences and social sciences have developed powerful tools for investigating environment–human relations and the responses that are required to avoid or mitigate problems. However, getting people to act in ways that lead to optimum, secure and sustainable lifestyles is a challenge: decision-makers have to win the support of people and other forces at work in modern 'enlightened democracies' and other countries, as well as promote appropriate technology and good governance (Barry, 1999).

1.1 The environment–human relationship

There is a long intellectual tradition of discussing the environment–human relationship. One element of the debate has been the effort to decipher whether nature is 'outside' social life, or constructed by social relationships. Irwin (2001: 23) argued that nature cannot be separated from culture, and that there 'is no single nature . . ., but rather a plurality of natures embedded in daily life'. He went on to argue that, 'Our ideas of the natural cannot be kept apart from the social concerns of our time' (2001: 59). Beck (1992) and Lash *et al.* (1996) also concluded that nature is not external to society. Consequently, current fears of an 'environmental crisis' can be seen to be linked to widespread social malaise, including loss of faith in centralized institutions and questioning of what constitutes 'progress'. But, decision-makers need to assess carefully the whole range of environmental threats and opportunities, as well as worrying about **development**-related environmental problems.

Glacken (1967: vii) observed that in

> . . . the history of western thought, men have persistently asked three questions concerning the habitable Earth and their relationship to it. [1] Is the Earth, which is obviously a fit environment for man and other organic life, a purposefully made creation? [2] Have its climates, its relief, the configuration of its continents influenced the moral and social nature of individuals, and have they had an influence on moulding the character and nature of human culture? [3] In his long tenure on Earth, in what manner has man changed it from its hypothetical pristine condition?

(The question numbers in square brackets in the above extract are my addition.) Botkin (1990: vii) posed similar questions about the fundamental relations between humans and nature.

- What is the character of nature if it is undisturbed by humans?
- How does nature constrain or influence human beings?
- How far is a human individual's behaviour, the shape of a culture and society due to environment, as opposed to nurture (i.e. life experiences, learning and free will)?
- How do human beings influence nature?
- What is the proper role for human beings in nature?

What do the terms 'nature' and 'environment' mean? It is easy to get mired in the literature trying to answer this; nevertheless, it is important because how people think affects how they perceive opportunities and threats, and the way they react. (Box 1.1 offers some definitions of environment and nature.) Glacken (1967: xi) provided a succinct and useful comment, that 'Man and Nature' is a useful shorthand for expressing a very complex relationship. (Nowadays, 'humans' would be substituted for 'man' and 'environment' for 'nature'.)

In western and westernized countries, the way people view nature has changed markedly over the last 400 years (Thomas, 1983), and especially since the 1960s. People in sixteenth-century England would have seen nature as something created by God to serve humanity, and they would have interpreted its features with reference to the Bible. In the Bible (Genesis 1: 26–28; and there is a similar passage in Genesis 9: 1–3) God tells Noah 'Be fruitful and multiply, and fill the earth and subdue it.' Roughly from the thirteenth until the mid-nineteenth centuries (AD), western Europe and the areas settled or influenced by it, are widely held to have adopted an anthropocentric Judeo-Christian outlook, which saw the world to have been Divinely created with resources, including all other organisms, existing to be used by humans to bring order to nature (see especially White, 1967a; 1967b; also Barry, 1999: 38–42). This worldview can be argued to have prompted exploitation, rather than stewardship, of nature. Elsewhere in the world, and more and more in the increasingly post-Christian and postmodern West, other perceptions hold sway, yet the same problems of exploitation and damage appear. Today many westerners have undergone or are undergoing a shift in their self-perception from seeing themselves as 'guardians' to 'abusers' of nature through overpopulation, pollution and other mismanagement – and with this the realization that they must live with the environmental and social consequences of a partially successful 'Promethean act', i.e. possession of many gifts and powers, but not necessarily enough wisdom to use them well (Arnold, 1996: 10).

By the late eighteenth century nature was seen as something to be exploited, and was increasingly interpreted in scientific terms. Mainly after the mid-1960s an awareness developed of the vulnerability, limits and need for stewardship, which manifested itself as 'environmentalism' (Forde,

1934; Arvill, 1967; Hewitt and Hare, 1973). Attitudes, even among one group of people, can change a great deal: in eighteenth-century England a middle-class person would have been likely to see undeveloped land as a wilderness, wasted or dangerous (the ideal – 'Eden' – was a tamed 'garden'); the equivalent person today would see wilderness as restorative, inherently healthy and might well donate money to Greenpeace or a similar body to protect it from development. At the start of the twenty-first century, nature is generally seen by the public and administrators to be stable, while natural scientists are increasingly aware it is in a state of flux or, at best, dynamic equilibrium.

In practice, there is much overlap in everyday usage between 'nature' and 'environment', both terms are widely seen to describe 'our planetary support system' (Botkin, 1990: 8). 'Nature' as a concept has strong links with birthplace and origins – milieu or cosmos might better be used to mean local conditions or surroundings that affect people; 'environment' implies the sum total conditions generated by a given interaction between organic and physical factors.

Humans interrelate with nature via culture. Today's environment is more and more a 'human construct', little is wholly natural and the influence of people is increasing – even the Earth's major biochemical cycles are being upset. With little contemplation of the consequences, humans are damaging the structure and function of nature. Nature functioned independently of humans for billions of years before *Homo sapiens* appeared. Some environmental change is being made more and more unpredictable because of human activity. It can be argued that humans have changed nature so much that now they alone can save and 'improve' things – although some supporters of the Gaia Hypothesis (see later in this chapter) would see this as hubris.

Nature may on the one hand mean all that nurtures humans, yet we are in constant battle with it – from deep within through our immune system,

---

Box 1.1 Some definitions of 'environment' and 'nature'

- Environment – limiting factors that determine quality of life, e.g. air, water, food, temperature, space, diseases.
- Environment – the 'sum total' of the conditions within which organisms live. It is the result of interaction between non-living (abiotic) physical and chemical, and – where present – living (biotic) parameters.
- Nature – all living things (except humans).
- Nature – essence of something.
- Nature – non-human surroundings.
- Nature – relating to one's place of birth.
- Nature – genetic tendency of living things, as opposed to their cultural expression.

Sources: Glacken, 1967: xi; Eisenberg, 1998: xx, 65; Barry, 1999: 12–18; and others.

which fights off infection and contamination – to external challenges like earthquakes or cosmic radiation from space. Many like to believe that humans can 'control' nature; yet, diseases that have exacted a terrible toll on human numbers in the past can still wreak havoc and natural disasters continue to be real, ongoing threats. Population growth, pollution and other misuses of nature are destroying biodiversity, wasting resources, and poisoning the air, sea and soil – reducing humankind's survival options. If humans do not control themselves better modern development may prove to be the mechanism of an ongoing mass extinction event comparable with past disasters recorded by palaeontologists. Humans also have to become more aware of their changeable environment.

When aware of environmental change we *interpret* it as threat, opportunity, irrelevant, attractive, unattractive and so on (often incorrectly). A people's attitude toward nature plays an important part in the formulation of their ethics – the rules and principles that shape and guide activities. Many authorities hold that in pre-scientific western cultures (and still in some societies), humans and nature were in harmonious relationship and this was seen to be the will of God (or the gods). The recognition of a divinely ordered, harmonious nature created for humans has been common, but not universal; in the first century BC, Titus Lucretius Carus (in *De Rerum Natura*) argued that it was not made for humans and was often chaotic. Since at least Classical Greek times some have viewed the Earth as an organism, sometimes hostile to humans; James Lovelock scientifically re-moulded such an organic view in the 1970s as the Gaia Hypothesis – nature kept in balance through a complex set of feedbacks between its organic and inorganic components (as discussed later in this chapter).

The predominant modern view is that natural 'resources' should be tamed and exploited. For many humans, contact with nature is now indirect. Early twenty-first-century westerners may recognize that a locality has 'charm' or 'character', and that wildlife has value, but they seldom go as far as their forebears, or some present-day indigenous peoples, to include concern for ancestors and the spirit world in their concept of nature. The dominant modern western (Eurocentric) attitude to nature is relatively materialistic, seldom values the aesthetic enough, and pays even less attention to spiritual matters or what is seen to be superstition.

Ethno-ecologists study various human groups' relationship with nature. Biologists have also explored the links: E.O. Wilson tried to establish if there is a genetic or evolutionary explanation for human interest in some aspects of the environment and non-human life (Wilson, 1984: 181–3; Kellert and Wilson, 1993), suggesting that scattered trees, which seem aesthetically attractive and comforting today, may once have offered real rewards – food, shade and refuge from wild animals.

Many countries have had (or once had) rules, laws and taboos that acted to control how people relate to nature (Fig. 1.1). In hunter-gatherer societies decisions about how to respond to environmental challenges and

Figure 1.1 Traditional net-fishing: lower Tocantins River (Pará), Brazil
Among the reasons for the declining stocks in many areas has been the breakdown
of the superstition, taboos and traditions that helped conserve fish in certain places
and at certain times.
Source: Author 1987.

opportunities are made by shamans, chiefs or village elders; in modern capi-
talist societies this is done by individuals in a position of power, the board
of a company, special interest groups or a meeting of government ministers.
In non-capitalist countries representatives of the state make decisions on
behalf of the people. In many nations today there are signs of empowerment, which should help a wider spectrum of citizenry to become involved
in decision-making. This includes indigenous peoples who are beginning to
have a voice and are consulted or even meaningfully involved in environmental decision-making.

Seeking purpose and order in their often changeable surroundings
prompted many of the ancients to treat the environment as something
sacred, and often to personify 'Nature' as a goddess of creation and life;
perhaps to balance this, some societies saw human culture as a more masculine side (Eisenberg, 1998: 65). One of the earliest clear indications that
people realized that the environment affects human fortunes is to be found
in Egypt, where for thousands of years Nile flood levels have been monitored so that advance warning can be had of likely agricultural productivity.
Some cave and rock-shelter paintings suggest that early humans (perhaps

before 40,000 BP) were aware of seasonal changes and developed some sort of calendar. During at least the last 4000 or 5000 years peoples in many parts of the world, including Latin America, Asia, Europe and Southeast Asia have gone to considerable trouble to observe and record natural phenomena, in some cases developing sufficient knowledge to accurately forecast complex astronomical events. In Europe and North America a 'Scientific Revolution' was under way by the mid-eighteenth century, established western humanistic rationalism and 'deprived nature of its sacred character' (McNeil, 2000: 328). To modern westerners the environment has become something external to humans, which can be understood through objective study. Presently, in spite of considerable progress, that study still needs to advance further to give enough knowledge to practise successful environmental management (McKibbin, 1990; Nisbet, 1991: 1).

Not all people have so practical and **utilitarian** an attitude toward nature; since the eighteenth century, in Europe and the USA, some have adopted various types of romantic stance. This was in part a reaction to the growth of industry and urbanization, pollution and centralized bureaucracy in the UK, western Europe, Russia and the USA (Emerson, 1954; Coburn, 1964). Romantics have sometimes helped stimulate others to undertake more practical measures: in the late nineteenth and twentieth centuries, encouraging the setting aside of conservation areas, and the development of environmental concern and green politics (Merchant, 1989: 21–7; Pratt *et al.*, 2000). Since the 1960s, growing numbers of people in western nations have started to advocate non-rational, often mystical, interpretations of nature. While any sincere concern for the environment is welcome, and it is refreshing to see appreciation of non-material values, it is difficult to see how any shift from rational and objective study, arguably humankind's greatest advance, will improve knowledge and contribute to better environmental management, let alone feed a crowded world.

Environment–human interactions are often very complex and changeable over time and from group to group, even among individuals; people experience technological, cultural, social and other non-physical variations as well. Fashions are particularly prone to vary – and this can mean that the way people react to a set of circumstances alters suddenly and markedly. Even in the eighteenth century, western Europeans still tended to blame adverse **weather** and natural disasters on God or witchcraft – indeed, some insurance companies are still inclined to do this if it reduces settlements.

Late eighteenth- and nineteenth-century natural scientists struggled to find order in nature and to determine universal laws. This 'Enlightenment', modernity, 'Age of Reason', or whatever one cares to term it, has shaped present attitudes; we owe much of our wellbeing to rationalism (together with adaptability it is one of the greatest gifts humans have), but the viewpoint is still widely held that nature behaves like a machine, and can be reliably modelled and, if need be, re-engineered to suit humans. Today, quite a few adopt a systems approach to environment–human matters and tend to see all change as interdependent interaction between physical, cultural, social and other factors. Reality is more complex and less sympathetic to human demands, and objective study may not give enough information to ensure warnings of all disasters and fully satisfactory environmental management – we should definitely not reject rationalism, but it must be augmented by ensuring that humans remain adaptable and resilient.

An organic (Gaian) view of environment–human relations has been offered, first in the seventeenth century, and more recently by James Lovelock and others. This presents nature as a self-regulating biogeophysical system of great complexity. Supporters argue that the Earth maintains a breathable atmosphere with about 21 per cent oxygen and an equitable temperature regime, unlike 'dead' sister planets, because bacteria, algae and higher plants have altered the original hydrogen, methane and ammonia gas mix. Acceptance of a Gaian viewpoint has implications for environmental management: if it is correct, exploitation and modification of nature will have to fit in, and clumsy re-engineering or maintenance is unlikely to succeed.

During the 1960s and 1970s a relatively small proportion of westerners were prompted, partly by photographs of the Earth beamed from space, and by a growing environmentalist lobby, to perceive the dependence of humans on a *finite* and sometimes delicate environment. Most early twenty-first-century humans are more out of touch with the environment than ever before, they are also likely to be less adaptive; yet they feel protected by what may prove to be a fragile bubble of technology and governance, and by the idea that lip-service to environmental concern is enough.

The term 'environment' was largely unknown to non-academics before the late 1960s, now it is in everyday use, frequently with little concern for its real meaning. It is for many essentially a relational concept, a social construction, whereby each user's views and experience determine what it means (the same can be said of the term 'nature') (Barry, 1999: 13). Although a lot of people use 'environment' simply to mean their surroundings, this is inadequate for exploring linkages between physical conditions and human fortunes. Nowadays, a common attitude is essentially existentialist, with adherents believing that humans have been thrown into a meaningless world. However, environment (or nature) is seldom just a meaningless physical backdrop to human affairs – people are a part, are influenced by it and increasingly affect it.

As discussed earlier, little of today's environment (if any) is natural and unmodified by humans. And, apart from a few artificial situations like spacecraft, nuclear submarines, strategic bunkers and to some extent the Biosphere 2 Experiment in the USA, all human life takes place within *one* global environment – there is no escape from physical conditions behind national boundaries or social class barriers. In addition, globalization is making people depend on resources from further afield and sensitive to wider-scale issues.

There are a variety of ways of ordering thinking about the environment–human relationship: one can adopt a utilitarian stance or one that is romantic, or adopt some other ethical–philosophical approach. Ideas about environment and nature are socially constructed and serve in different ways as instruments to aid authority, identity and defiance. So, one person's 'threatening wilderness' may be another's 'Eden' (Arnold, 1996: 3). There is growing realization that humans must act to manage the environment; those seeking to do so range from 'environmental managers', confident they can use science, technology and wisdom, to those who adopt a romantic or even mystical approach. Eisenberg (1998: 240) calls the romantics 'fetishers' – environmentalists who at their most extreme wish to abandon modernity and its technology to return to an unspoilt 'Eden' where there is minimal human impact. However, as argued earlier, this is not a painless option; as Eisenberg (1998: 73) observed, 'Never before have so many tried to live in Arcadia!' With a human population of over 6500 million, a low-tech Eden would be suicide for most people – future survival lies in developing and 'tuning' what we have, not in rejecting science and technology.

1.2 Eden, Arcadia, Utopia

Many peoples have legends about having lost a past state in which humans were close to the point of their creation and in harmony with nature, the gods or God. For example: the Judeo-Christian-Islamic story of the expulsion from Eden; the Hellenic Greek (Epicurean) myth of a past Golden Age in Arcadia (a region of Greece south of the Peloponnese is still known as Arcadia); or Australian Aborigines' folk tales of a primordial Dreamtime.

The Eden and Arcadia myths possibly originate from a point in human history when we began to see ourselves as separate from nature (in Palaeolithic times?) (Glacken, 1967: x). It may relate to a marked alteration of lifestyle as the Quaternary environment alternated from interglacial warmth to glacial chill or when the last ice age gave way to postglacial conditions. Eisenberg (1998: xv) suggested that these feelings of having had something go wrong between humans and nature were indicative of an awareness that there is currently some sort of discord (Dystopia). Another possibility is that there are memories of supposed easier hunter-gathering before conditions changed during the Holocene or earlier (see Table 2.1), prompting the development of sedentary agriculture and settlement in villages, towns and cities. Fagan (1990: 61) suggested that 'Eden' was the savannahs of sub-Saharan Africa, which offered a fairly undemanding 'nursery' environment for hominid-human evolution, where Quaternary climatic fluctuations were relatively gentle and food was abundant. Anthropologists like Hugh Brody (2000) have argued that the transition from pre-agricultural to farming livelihoods condemned people to 'become exiles bound to move over the earth, struggling to survive'. Effectively, the adoption of farming has made people more vulnerable to nature, and has often meant harder work, poorer standards of living, narrower horizons and new diseases. Another possibility is that civilizations developed much earlier than is generally accepted, during the last ice age, in the tropics and sub-tropics on land now submerged, and tenuous memories have survived of the flooding (see the discussion of world flood myths in Chapter 2).

Utopia is the dream of a perfect future. The idea that humans might regain something approaching Eden (during life) through industry, skill, justice and wisdom was seldom promoted in Europe before the sixteenth century (to do so would probably have invited execution for blasphemy). Then, in AD 1516, Sir Thomas More published *Utopia*, a book that discusses what today might be termed 'development'. (Some are puzzled that More, a devout Catholic, offered a blueprint for improved earthly society – perhaps it is also an appeal to people to mend their ways.)

Since the eighteenth century improved understanding of natural history, technological advances and exploration of the world, has meant that some peoples have improved their physical wellbeing and others have been encouraged to dream of further development. This development has not been universally welcomed. During the eighteenth and nineteenth centuries Europeans encountered novel environments and peoples, like those of Tahiti and Fiji, at a time when there was widespread concern about urbanization, industrialization and social degeneration; these contacts provoked romantics like Jean-Jacques Rousseau (1711–78) to compare the new islands to Eden and to extol the virtues of 'the noble savages'. Polynesians, North American Indians and Australian Aborigines were supposedly sympathetic to nature – 'learn from them' was the message. However, a good deal of this supposed rapport with nature was more imagined than real – an 'Arcadian myth'; for example, prehistoric hunter-gatherers seem to have managed to decimate game animals enough to cause extinction in many parts of the world; and, worldwide, early man considerably modified natural vegetation with fire and stone axes. Anthropologists may have been too eager to recognize indigenous peoples' adaptations to the environment – rash generalizations have been based on one or two case studies or misconceptions (Lewis, 1992: 49–51). Humans have always been opportunists who sometimes fail to exercise sufficient stewardship of their environment and its resources.

While some romantics saw tropical nature as Eden, it was more common for eighteenth- and nineteenth-century Europeans to treat indigenous peoples as a non-human part of the environment; a convenient assumption, because the land could then be legally treated as *terra nullis* – 'uninhabited', and claimed with little or no payment or feeling of guilt.

The optimistic name of 'Vinland' applied by Norse adventurers to North America, and some of today's interest in science fiction and space exploration, has similar roots in dreams of Utopia. By the mid-eighteenth century some intellectuals and dissenters were actively exploring routes to Utopia –

for example, the founding of independent America or the French Revolution. Twenty-first-century developers mainly see the path via technology and 'civilization': 'The more mature and civilized a society becomes, the less it lies in nature's thrall: indeed the march of civilization is precisely its ability to rise above narrow environmental constraints' (Arnold, 1996: 10). Deep-green environmentalists, neo-pagan and New Age groups seek to pursue Utopia through moral and social change and, at best, limited adoption of technology.

1.3 The Gaia Hypothesis

The Gaia Hypothesis, developed by James Lovelock since 1969, has, in spite of fierce debate, increasingly gained acceptance (Lovelock, 1979; 1988; 1992). Similar 'organic' worldviews had been proposed earlier; for example, by James Hutton in the early nineteenth century and Pierre Teilhard de Chardin in the 1940s. Lovelock was prompted to put forward his hypothesis when working on instrumentation for the NASA *Viking* Mars landers in the early 1970s. Mars and Venus, he noted, were in chemical equilibrium, whereas the Earth was not; living organisms, he surmised, were the reason the Earth had maintained atmospheric oxygen at between 15 per cent and 22 per cent of the total gas mix, roughly similar to present levels throughout at least the last 600 million years (and perhaps for over 2000 million years). Apart from a few relatively brief 'icehouse' and 'greenhouse' phases, life has also helped keep temperatures more or less constant. This thermoregulation has been maintained, even though early in the Earth's history (*c.* 4.6 **billion** years BP) the Sun is believed to have been 30 per cent to 40 per cent weaker than it is today. Also, there have been disasters that disrupted the terrestrial environment but recovery has occurred – probably through organic activity. The control has been marked – when the Earth was young it had an atmosphere rich in carbon dioxide, possibly as much as 60 times as dense as today. Subsequently, much of the carbon dioxide has been locked up in carbonate rocks, with organisms playing a key role in carbon sequestration.

So, life has apparently been active, not passive, reacting to environmental change through a web of checks and balances to maintain a homeostasis (which it helped establish). Life did not simply evolve to fit the planet's physical conditions, it has also changed the environment to suit its needs. The suggestion that the Earth behaves like an organism, evolving mechanisms that allow self-regulation, also implies that humans need nature more than nature needs humans – if we do not 'fit in' we may well suffer. Lovelock's supporters argue that to understand such a system, and to work within its limits, demands a holistic, Gaian approach.

The Gaian system is not seen by Lovelock and mainstream adherents to be in any way a conscious process, nor is there implied supernatural design or purpose. Most of those supporting the Gaia Hypothesis accept that the

Earth's biota affects physical conditions; although a few go further and claim co-evolution of physical and biological; and there are a minority who suggest that life or some supernatural force 'intentionally' alters physical conditions.

1.4 Nature and culture

The environment–human interrelationship is two-way and complex: there are often positive and negative feedbacks (i.e. magnifying or counteracting changes respectively), and direct and indirect interactions between physical, biological, behavioural and cultural factors. Nature and culture are so entangled that it is difficult to separate them (Dubos, 1973). Humans adapt to changes through both biological evolution and learned behaviour, and the path taken is often erratic and difficult to predict. Most people belong to a society and a culture that are changeable, and the variation can affect vulnerability to environmental factors and ability to grasp opportunities and adapt (Bennett, 1976a). Culture and technology are traits like the teeth of a beaver, adaptive mechanisms that assist humans to function within an ecosystem. With the environment always capable of presenting challenges human adaptability is important and depends upon given social and cultural qualities. In 1998, I and a colleague visited valleys in the High Atlas Mountains (Morocco) where neighbouring villages fared very differently in the face of similar environmental conditions. **Social capital** (see Glossary) was not of the same strength in each village, and this affected how the villagers coped with similar environmental and socio-economic challenges (Barrow and Hicham, 2000). Early settlers in the Americas would probably have perished in the first severe winter if indigenous peoples had not shared know-how, helping the newcomers to adapt.

Culture (see Glossary) helps to determine behaviour and often functions to maintain a community in equilibrium with its environment. Culture is not a barrier separating humans from environment but a linkage. However, there are situations where cultural traits may continue even when they do not 'fit' – as Noël Coward pointed out in his song 'Mad Dogs and Englishmen'. Some communities may develop cultural traits that reduce their long term chances of survival: greed and consumerism in modern western societies are examples. Cultural change can be gradual or sometimes sudden and marked, and it can be triggered from outside and within a group. The forces altering culture include: innovation, invasion, political change, shifts in fashion, changed attitudes (and many other 'human' factors), and 'environmental' factors like natural disasters or more gradual physical changes.

There has been a long-running debate about cultural evolution. Cultural evolution is complex and in some ways resembles biological evolution. The important biological difference between cultural and (genetic) evolution is that individuals producing and transmitting culture are likely to be more biased and selective in what they transmit to future generations than nature is with genes. Julian Steward (1955) argued that, over time, people tend to organize themselves in similar ways to exploit comparable environments; consequently there is a sort of parallel evolution of cultures in different parts of the world. The orthodox view was that culture evolved, leading to more complex societies that used resources more efficiently, were more adaptable and enjoyed more security (Bodley, 2002). More recently, some theorists have argued that small-scale cultures (such as groups of hunter-gatherers) might actually be more adaptive and provide a better 'quality of life' than modern 'more evolved' cultures. Considerable human cultural evolution has taken place in the last 100,000 years (*c.* 5000 generations); over this span of time there is not much chance for physiological change to occur (Haberle and Chepstow Lusty, 2000). Understanding the environment–culture relationship is vital if nature is to be better protected and human welfare improved, and a growing number of researchers have addressed these issues (Croll and Parkin, 1992; Leach and Mearns, 1996).

Catastrophism

There had been a few supporters of catastrophism before the nineteenth century (for example, the English writer Francis Shakelton published such views in *A Blazying Starre* in 1580); however, few thought of nature undergoing much change, either gradual or sudden. The general attitude in Europe at the start of the 'Age of Reason' (mid-seventeenth century) was that the world had existed only for a limited time since the Creation and had been made 'perfect' in its current form. In 1658 the Archbishop of Armagh (Ulster, UK), James Usher, estimated from Biblical genealogical sources that the Earth had been formed at 9 am on 23 October in the year 4004 BC. Establishing so short a span for history reinforced the prevalent view in western Europe that the world had been shaped by only one catastrophic 'Noah's flood'. The terrible Lisbon earthquake of 1755, and associated tsunamis, showed that nature could undergo change, sometimes sudden and violent, and this caused widespread culture shock.

The idea that the Earth had not developed much since its formation a few thousand years in the past gradually came under pressure. Debatably, it was Nicolaus Steno (1638–87) who first argued that sedimentary rocks were deposited over time and had not all been made at the same moment of Creation, and Giovani Arduino (1713–95) who suggested the concept of superposition – that lower rocks are older than those above them (provided that there has been no displacement or overfolding). The debate came into the open and intensified after Charles Lyell (1797–1875) published *Principles of Geology* (three volumes in 1830, 1831 and 1833). This influential work, together with that of other eighteenth- and nineteenth-century geologists and natural historians, particularly William Smith (1769–1834), James Hutton

(1726–97), Georges-Louis Leclerc (the Compte de Buffon) (1707–88), Baron Georges Cuvier (1769–1832) and Jean-Baptiste Lamark (1744–1829), helped establish that the Earth was immensely older than a few thousand years. A sea change was occurring: the shift to accepting that the Earth and its organisms had gradually evolved over many hundreds of millions of years, and that life was still changing and not something created in its current form (there are still many people who do not accept this view of nature). There also dawned the realization that processes in operation today have in the main shaped things in the past (the concept of **uniformitarianism** – pioneered by James Hutton in 1785 and supported by Lyell, Cuvier, Louis Agassiz, William Buckland and others). With a huge span of time for slow processes to shape organisms and the physical environment there was less need to invoke catastrophes to explain the Earth's features – and there was time for biological evolution to operate (though at that point nobody knew how).

The environment's changes became obvious to many scientists by the early nineteenth century: sea-levels had altered, land had been uplifted, marine fossils outcropped on mountains, remains of tropical organisms occurred in regions that were far from the Equator. All this suggested eustatic, tectonic and **climate** change (the idea of continental drift was not widely accepted until the late 1960s). In 1837 Louis Agassiz, convinced by two Swiss geologists, promoted the idea that high latitude and alpine ice sheets had expanded and contracted far beyond their then limits during relatively recent geological history.

In 1859 Charles Darwin published *The Origin of Species by Means of Natural Selection*; one of the key concepts he put forward was that individuals with favourable characteristics tend to survive to reproduce – a mechanism for biological evolution. Thus organisms evolve to 'fit' their environment (although it is now clear that occasional catastrophic events may overwhelm even the 'fittest' species, possibly allowing the lucky and less 'fit' to survive and flourish). Darwin's ideas helped prompt the view that to some degree the environment determines what shape organisms take. How far this is true is still hotly contested – the 'nature versus nurture' debate – seeking to establish to what degree human behaviour and fortunes are determined by genetic and biological factors, upbringing and learning, or environment (Williams and Ceci, 1999). Darwin's theory of natural selection suggests that if a species becomes too specialized it may be unable to adapt or move if there is an unfavourable change in environment. In the past humans have tended to be generalists; are modern humans and their societies too specialized for their own long-term good?

Neo-catastrophism

Neo-catastrophism is often associated with Otto Schindewolf (1896–1971), who published a seminal paper on the subject in 1963; he used the term in relation to mass extinction events and global-scale causation. In the period since the 1950s there has been a growing awareness among natural scientists that there are catastrophes superimposed on the 'norm' of steady conditions: the former may prompt bursts of rapid evolution and the latter support more gradual biological evolution and physical change for most of the time (this is often referred to as 'punctuated equilibrium'). Even if the present is the key to the past (the uniformitarian view) humankind has too limited recall to be sure of what happened before a few thousand years ago. Derek Ager (1993: xvii) observed that 'improbable events become probable over a sufficiently long time span'. Hence prediction of the future is difficult from historical hindsight. Palaeoecology, geophysics and other sciences have shown that there have been infrequent but marked disasters in the past. Again, Ager (1993: xix) vividly made the point: 'The world's history is like that of a wartime soldier – long periods of boredom and short episodes of terror.'

There is popular interest in catastrophes – they make for quite a successful film genre – but governments have been reluctant to consider such threats seriously. Efforts to identify and avoid or mitigate some frequently recurrent disasters have been undertaken: the British tried to forecast and seek to prepare for monsoon failures in India during colonial times, and in most western countries, the former USSR and the People's Republic of China, civil defence contingency planning was well supported during the Cold War. Sensationalist catastrophists, like Emannual Velakovsky writing in the 1950s, who had little provable foundation for their claims helped to give the study of catastrophic events a bad name (Velakovsky, 1950; 1952; 1955). Much of the writings of environmentalists in the 1960s and 1970s, notably that of neo-**Malthusians** is essentially catastrophist.

Palaeoecological data about catastrophic events has expanded; for example, there is now good evidence for vast, short-lived, but highly erosive floods during the melting of the North American ice cover between 20,000 and 10,000 years ago. Some of these floods carved the Channelled Scablands of Washington State (approximately 13,000 BP) when a meltwater lake (Lake Missoula) in western Montana suddenly drained, it did so repeatedly with catastrophic effect. In perhaps as little as 48 hours the flood eroded gorges and transported boulders as large as 300 tonnes, leaving gravel deposits across much of Montana, Idaho, Wisconsin and Oregon, and probably caused significant sea-level change on a global scale (see http://www.uidaho.edu/igs/rafi/iafidese.html, accessed 25 May 2000).

Some scientists and activists, but fewer administrators, have begun to take an interest in the possibility of early warning and avoidance of rare, but major catastrophes like asteroid impacts. In a number of countries the threat of earthquakes, volcanoes and tsunamis is taken seriously. Events like the Mount St Helens and Montserrat eruptions may have helped stimulate interest, but efforts are too uncommon and too limited in scale.

Neo-catastrophism has, since the 1960s, been used to describe one of two opposing viewpoints about the consequences of global environmental limits and environment–human relations; the opposite is the cornucopian viewpoint. The neo-catastrophist (Cassandran, neo-Malthusian or pessimistic) view argues that there are environmental limits to human activity that cannot be exceeded without disaster following; indeed, some would claim that development has already overshot and passed these limits. The solution is to limit growth and ensure that development is environmentally sound. The cornucopian (optimistic) view claims that human ingenuity and **adaption** allow development to bend the limits (Meadows *et al.*, 1972; 1992; Goudie, 2002: 141–4).

1.5 Human ecology and historical ecology

Discrete fields of study have evolved to investigate environment–human linkages. By the early 1920s urban geographers at Chicago University were stimulated to develop human **ecology** by the work of botanists like F.E. Clements (1874–1945), who studied plant community succession (phytosociology), and by W.M. Davis (1885–1934), who tried to establish the idea that there was a cycle of landscape and geomorphological development. Both argued that it was possible through the study of successive stages to predict future outcomes and that the evolution led to relatively stable steady-state landscapes or climax communities of plants. Human ecology tries to apply ecological ideas to humans (Eyre and Jones, 1966). Before 1945 human ecologists saw culture as the 'steady-state' equivalent to a climax vegetation (Clements, 1916; Barrows, 1923; Park and Burgess, 1924). Historical ecology overlaps with human ecology; originally developed by anthropologists, its practitioners use diachronic research to help them understand human development (Crumley, 1994).

1.6 Environmental determinism

'Determinism' is the theory that all human actions are caused predominantly by preceding events and not by the exercise of will. The idea that environmental factors like prevailing temperature and humidity, climate change and disease have a determining effect on human history is not new: on the Aegean island of Kos around 5 BC the Greek philosopher Hippocrates reputedly observed (in the second half of a work attributed to him: *Airs, Waters, Places*) 'the constitutions and habits of people follow the nature of the land where they live'. The Roman philosopher Vitruvius (*c.* 40 BC) noted that 'peoples and cultures are as they are because of environmental conditions'.

Few dispute that building location and style tend to reflect physical conditions; houses are often sited close to springs or streams, and to catch the sun, avoid frost hollows or risk of floods, and so on. People often argue that decor, furnishing and landscaping of dwellings affect behaviour. Interestingly, earthquakes, volcanic eruption and tsunami threats seem to have much less effect on construction or siting. Humans choose what they are willing to be influenced by (Sjoberg, 1987).

Philosophers, theologists and scientists have long been interested in exploring the extent to which human behaviour is determined by supernatural forces, environment, biological or cultural factors, or whether individuals do actually have 'free will' (self-determinism). Those supporting environmental determinism (co-evolution or causal correlation) argue that human behaviour is to a significant extent decided by nature. Adams (1992: 12) pointed out that environmental determinism is attractive because it offers a coherent pattern of causation so that human history becomes more than just a series of random events.

By the mid-nineteenth century it seemed that most early ('primitive') cultural hearths were at low-needed latitude, but that progress had occurred mainly at mid-latitude. Also, some societies were clearly much affected by nature, notably the Egyptian civilization, which depended on adequate flows from a variable River Nile. In the seventeenth century the French philosopher the Baron de Montesquieu (Charles Secondat) suggested that climate affected physiology, which in turn determined psychology – so that even religion might reflect physical conditions. In the late eighteenth century it was widely held that westerners, not 'natives', had a mission to remove weeds and vermin, and to order nature (especially in the New World) to recreate something like Eden. Also, in the late eighteenth century the writings of Thomas Malthus (1766–1834) prompted a pessimistic deterministic view of nature; his first *Essay on the Principle of Population* in 1798 pointed out that human population growth would, unless constrained by war, pestilence or self-restraint, exceed environmental limits. This work had a profound effect on many, including Charles Darwin, and probably on those in power who were influenced to withhold aid during the Irish Potato Famine of the mid-1840s. A strange example of eighteenth- and nineteenth-century determinism is phrenology, proponents of which claimed to be able to tell if an individual had inborn criminal or other behavioural traits from the shape of their skull (many police forces and blue-chip companies still use assessment of handwriting to make judgements about suspects and job applicants).

Criticism of crude environmental determinism dates back at least to the eighteenth century. Voltaire is quoted as saying 'much of the behaviour involved in interactions with the environment is learned' (Vayda, 1969: xi) and 'climate has some influence, government a hundred times more; religion and government combined, more still' (in Commentaire sur Quelques Principales Maximes de l'espirit des Lois', *Oeuvres*, ed. Beuchot, Vol. 50, pp. 132–3; cited in Glacken, 1967: 582). Presently, the majority of academics reject environmental determinism as flawed, rooted in primitive Greek

and Roman philosophy, and heavily seasoned with racial and western imperialist ideas (Jones *et al.*, 1999: 137). Certainly, environmental determinism
has a 'shady past': between the 1870s and 1920s extreme and 'crude'
versions were promoted by a number of European and American intellectuals – notably Richter, Kant, Ritter, Ratzel, List and Churchill (Ratzel,
1889; Pepper, 1984: 111–12). Many environmental determinists and Social
Darwinists saw Europeans as superior and evolved by natural selection,
other races were at less advanced evolutionary stages.

Ellen Churchill Semple (1863–1932) argued that the environment–
human relationship was such that parameters like climate directly and
indirectly influenced and substantially controlled behaviour, culture, society
and development (Semple, 1911). This simplistic determinism held that
environment (and race) largely shaped development; put crudely, temperate
climates ensured that Caucasian peoples were intelligent and industrious,
and destined to achieve great things; whereas tropical conditions, especially
when combined with non-Caucasian race, were likely to lead to underdevelopment; and very cold conditions lead to morose and aggressive
characteristics (see Boxes 1.2 and 1.3).

Before Charles Darwin published *The Origin of Species by Means of
Natural Selection* there had been political thinkers and natural philosophers
promoting ideas that broadly argued the 'survival of the fittest'. In the
nineteenth century the English philosopher Herbert Spencer (1820–1903)
developed the concept of Social Darwinism. As Gould (2000: 257–67)
noted, this was roughly a decade before Charles Darwin published, and so
would better be called 'Social Spencerism'. In his book *Social Statistics*
(1850), Spencer argued that genetic or biological factors significantly
determined social or sexual inequalities; individuals, groups and races were
subject to evolution, class and racial differences were 'natural' and progress
was through 'the survival of the fittest' (Spencer's, not Darwin's, expression).
Social Darwinism still has supporters and has been blamed for delaying

---

**Box 1.2 A simplified presentation of the discredited views of early
twentieth-century environmental determinists like Huntington (1915)**

Table 1.1 The discredited views of early-twentieth century environmental determinists (simplified)

Determinants	'Level of sophistication' 1=highest 4=lowest
'strong' race + favourable climate	1*
'strong' race + less favourable climate	2
'weak' race + favourable climate	3
'weak' race + less favourable climate	4

Note: *Assumed by determinists to be Caucasian + temperate environment.

Box 1.3 Some examples of 'traditional environmental determinism'

Various cultures have long argued that environmental factors affect human wellbeing and efficiency: the vapours from swamps; periodic weather conditions, especially unfavourable winds such as:

- the Scirocco – a hot, damp wind from North Africa that has long been held to cause intemperance and violent behaviour in southern Italy
- the Mistral – a violent, very cold north or northwesterly wind that is supposed to cause depression in parts of southern France
- the Bora – a wind of the Adriatic and Balkans with a similar supposed effect to the Mistral
- the Levanter – a cool and damp wind, which strikes Gibraltar from the east and is reputed to cause lassitude.

Note: A northerly wind comes *from* the north; a southerly, *from* the south, and so on; but northerly ocean currents flow *to* the north, and so on.

---

improvement of social welfare and workers' rights and measures to reduce inequality in the USA and elsewhere as late as the 1960s. Supporters claimed that humans were at the top of an evolutionary hierarchy, and that competition, self-interest and struggle, rather than cooperation and mutual aid, were natural and justifiable ways to behave, and that the group best able to adapt to its environment should become dominant. Selective breeding and suppressing or discouraging the 'weak' and 'undesirable' – eugenics – were by the 1920s supported by most Social Darwinists as a way to improve a particular human group's genetics, and so better its chance for long-term survival and its achievements. Social Darwinism and eugenics were embraced in Nazi Germany (where the 'blood and soil doctrine' was founded on a belief that race, environment and culture were intimately related), and less extreme forms were still supported in parts of the USA and some other countries in the 1960s and later (Irwin, 2001: 9). One of the key faults of crude environmental determinism is that people are unlikely to have a fixed response to an environmental challenge or opportunity – they can choose a course of action that, over time and from group to group, can be very different.

Researchers have paid much attention to the impacts of climate on human history. Climatic determinists and historical climatologists overlap some of their interests, although the latter are more concerned with establishing past climate conditions than interpreting human impacts, and rely much more on empirical data and less on often speculative reconstruction. Synoptic climatologists use recorded observations to try and establish past climates. Huntington, in the first decades of the twentieth century, tried to correlate an impressive collection of historical and climatic information with human health, energy and mental ability, publishing the results in a number of widely read works (including *Civilization and Climate*; *The Pulse of Asia; Mainsprings of Civilization; World Power and Evolution* and *The Character of Race*). Huntington's message was that civilizations' rise and fall, mainly through climatic causation. He also tried to map 'human energy' and 'creativity', resulting in the conclusion (like Semple's) that these were highest in temperate climes and lowest in the humid tropics: 'All depends on the two fundamental conditions of race or inheritance, and place or climate' (1915: 22). He also felt that humankind's 'physical and mental energy and his moral character reach their highest development only in a few strictly limited areas' (1915: 285). While not as openly racist as some determinists, these views fitted nicely with the Eurocentric and imperialistic outlook of the time (Freyre, 1959; Chappell, 1970; Martin, 1973; Alatas, 1977).

Gould (2000: 212–14) scoffed at the idea that European domination was foreordained because of environmental determination (and certainly not race); while accepting that nature does influence *on a broad scale*, he argued that if history were re-run it would probably not give similar results, because a slight shift in only one component of a hugely complex chain of causation would cause a different outcome. Although much of crude environmental determinism and 'naive anthropomorphism' has been discarded, some ecologists still make use of environmental zones to help understand situations. One of the best known of these approaches is the World Life Zone System (Holdridge, 1967). In reality there is so much difference from site to site in microclimate, etc., that such zoning is only useful as a coarse tool. If climate does affect human choices and fortunes, it may not necessarily be just the normal 'average' conditions but rather brief and easily missed events like a single frost or one storm. There are also likely to be situations where more than one factor comes into play (cumulative impacts) – for example, a cold winter alone may not have a significant effect, whereas cold plus wet, or cold and windy, or unseasonable conditions might.

Since the 1920s there have been a number of 'rational' environmental determinists and environmental historians prepared to argue that nature is a significant element in human affairs (Markham, 1942; Claiborne, 1970). One influential group of French academics – the 'Annalistes', which included Marc Bloch (1939), Lucien Febvre and Emmanuel Le Roy Ladurie, established the deterministic journal *Annales* in 1929. Toynbee (1934; 1946; 1976), discarded race as a determinant, and focused on environmental controls. Even so, Claudio Vita-Finzi (in Slack, 1999) urged caution, describing Toynbee's approach as 'climatic determinism in a velvet glove'. Determinism has influenced anthropologists and post-1960s environmentalists. Turner (1962) explored the significance of the 'open' frontier on the development of the USA before the 1890s. US institutions and attitudes, he argued, were determined by expansion enabled by an open frontier, whereas those in the Old World had been shaped by closed boundaries, limited opportunities and crowding. Although Turner's views have been criticized, a number of environmentalists since the 1960s have stressed that 'closing frontiers', limits to growth and a finite 'spaceship Earth' ultimately largely determine human fortunes.

While crude environmental determinism has been discredited (see Box 1.4), some have continued to comment on the apparent correlation between location in the tropics and underdevelopment (Biswas, 1979). J.K. Galbraith observed that if 'one marks off a belt a couple of thousand miles in width encircling the Earth at the Equator ... [one] ... finds within it *no* developed countries' (1951: 693). *If* there is a link between tropical environment and underdevelopment, then a very large part of humanity is affected – over 40 per cent of the Earth's population (Lee, 1957: vii; Harrison, 1979). The causation is not proven and ignores archaeological evidence of civilizations in the tropics in the past and some present-day city-states; also, recent tropical underdevelopment might equally be due to colonialism, isolation from Europe and North America, or other non-environmental reasons, as much as leached soils, high temperatures, 'tropical' diseases or unfavourable precipitation. Recent studies of the possible impact of global warming (the 'greenhouse effect') have focused on the possibility that the tropics may experience impacts that differ from those of higher latitudes – for example, tropical crops may have a form of photosynthesis that will be impaired by elevated temperatures and carbon dioxide levels, whereas tropical weeds may flourish, and in higher latitudes the opposite might occur.

Box 1.4 Some criticisms of environmental determinism

- It makes loose assumptions about correlations, and correlation does not *prove* causation (neither does holistic generalization)!
- It makes selective use of evidence.
- Correlations are often attempted using relative dating (e.g. radiocarbon or palynology); without *precise* dates it is difficult to exactly match often short-lived events.
- People can rebel against nature and may even alter it.
- Different cultures have sometimes evolved in the same environment, which seems to contradict environmental control.
- It has tended to be an ethnocentric (Eurocentric) concept.
- Few environments today are natural; most (if not all) are to varying degree modified by humans.
- Assumptions are usually too simplistic to adequately cover the complexity and subtlety of environment–social, cultural, and technological issues.
- Environmental determinism is likely to underestimate the importance of factors like human greed.
- Environment–human relations are mainly via complex chains of causation, not simple and predictable mixes of variables.

Source: Compiled by the author.

1.7 Environmental possibilism

Environmental possibilism was developed by the French geographer Paul Vidal de la Blanche (1845–1918), and further refined by Febvre (1924), as a refutation of environmental determinism. Supporters of environmental possibilism hold that nature or environment must play some role, there *are* absolute limits of human tolerance to cold, pressure, heat, access to water, etc., and these *limit the range of possibilities that culture and society can work upon* (that is, they limit the alternatives available but do not determine which will be chosen) (Vidal de la Blanche, 1918). Vidal de la Blanche pointed out that humans can selectively respond to any factor in a number of ways. How people react and develop is a function of the choices they make in response to their environment. These choices are the manifestation of their culture. So, possibilism suggests that humans' response is partly a function of their physical environment and partly of their own free will. Put simply, possibilism holds that the natural physical environment and resources are both a facilitating and a limiting frame for human action – *the environment may limit but doesn't directly control* – so humans are masters of the environment and not vice versa. This makes it possible for different cultures with varying adaptations to exist in the same physical environment at once, in succession or at different locations within the same environment.

Possibilism is quite similar to self-determinism, a philosophical or theological viewpoint which accepts that hereditary, environment and other factors *affect* an individual's behaviour. However, self-determinism denies that these factors *determine* behaviour. Those who have accepted a possibilist viewpoint include Thomas Aquinas (centuries before Vidal de la Blanche) and, more recently, C.S. Lewis. Possibilism proved attractive to Stalinists in the USSR (Ellen, 1982: 21–51) and it is still widely favoured today in the West. Fieldwork amongst the Inuit by Franz Boaz in the early twentieth century is often cited in support of possibilism. Clifford Geertz (1963) suggested that people could boldly re-work nature (citing the example of wet rice terraces in Java), or embark on a canny mimicry of nature (various forms of shifting cultivation), and that some colonial regimes and international market situations could constrain economic development, forcing peasant agriculture back into traditional ways (an involution of the cultural–ecological process).

There are problems with possibilism: to say that the environment just limits development is too simplistic – it may also prompt it. Some adopt a slightly more intermediate position, which recognizes that humans can *improve on nature*, yet accepts that environmental limitations exist and are especially likely to be apparent if there is unwise resource use.

1.8 Particularism

Particularism is the view that cultures, environments and histories are so varied that generalization is hazardous. So it is sensible just to view environmental conditions as *constraints*. Environment can be an influence, but culture and technology vary and can be independent of it (Forde, 1934).

1.9 Cultural ecology

Cultural ecology is a school of ecological anthropology that some feel might serve as a better alternative than environmental determinism or environmental possibilism. It has been defined as the 'study of the adaptive process by which human societies and cultures adjust through subsistence patterns to a given environment' (Ellen, 1982: 53). Cultural ecologists generally make use of the concept of carrying capacity and see culture as something through which humans adapt to environmental conditions. The approach, which became popular with anthropologists in the 1950s and 1960s, was in part established by the ecological anthropologist Julian Steward (1902–72) in the late 1930s and 1940s. Steward argued that cultures interact with their environmental settings through a process of **adaptation** (often prompted by technological innovation), leading to cultural change (Steward, 1955; Harris, 1965; 1971; 1974; 1992; Netting, 1977).

Cultural ecology adopts a local livelihood focus, explaining how human populations adapt to their environment, concentrating on a key 'core' of cultural factors involved. It explores how religion, family values, technology, politics, kinship and so on can affect subsistence, and thereby adaptation to environment. Cultural ecologists recognize adaptive responses to similar environments that can give rise to cross-cultural similarities. Supporters of cultural ecology hold that:

- the environment offers challenges and opportunities, and that culture can adapt to these (not the simple physical limits of environmental determinism)
- through adaptation a culture can change to best 'fit' an environment
- the existing cultural features affect how the environment influences humans
- there could be more than one possible outcome in similar environmental circumstances.

Steward's ideas on cultural ecology were developed and modified by Marvin Harris (1992) who explored the struggle of society to exploit nature. One of his studies argues that the taboo on eating beef in India evolved as a rational reaction to environmental conditions (cultural materialism).

Dissatisfaction with cultural ecology since the 1960s has led ecological anthropologists to develop historical ecology, ecosystem model ethno-ecology and ecological functionalism (for information on ecological anthropology, including websites on cultural anthropology, see http://www.as.ua.edu/antFaculty/murphy/ecologic.htm, accessed 25 May 2000; Kottak, 1989; Little, 1983). Ecological functionalism is a school of ecological anthropology that takes more account of feedback processes in the environment–human interrelationship, and allows more for subtle factors like rituals (for a short review of this see http:/courses.washington.edu/anth457/hist-lec.htm, accessed 25 May 2000). There is currently a trend amongst ecological anthropologists and some allied disciplines, such as human geography, to adopt a **political ecology** approach (see Glossary).

While discussing the degree to which culture may determine human fortunes it is appropriate to consider fatalism. Individuals or groups may fail to grasp opportunities or react in a beneficial way to a challenge, even when they are aware of the situation and may have the necessary skills and resources. This is because they feel powerless, disheartened, or hold political, philosophical or religious views that prevent them from acting. Similar inertia can appear when people are faced with a choice of spending resources on what they feel is a 'remote' threat or something not directly affecting them.

New-environmental determinism (neo-determinism or evolved determinism)

From the 1940s to the 1990s, environmental explanations for abandonment of cities and the decline of past civilizations have generally been dismissed in favour of blaming human causes. Some favour a postmodernist focus on power relations, social conflict, elite conspiracies, gender inequalities and so on. The failure by many to consider environmental causation is unwise (Jones *et al.*, 1999), and since the late 1980s, in fields like archaeology and agricultural development, there has been some interest in cautious forms of environmental determinism. For example, Carl Sauer (1952) argued that plant domestication was unlikely to have happened if there had been *sudden* dire need – people would starve before they could experiment – so new crops resulted from more gradual environmental and cultural stimulus, and prompting, with the innovators having enough food security to be able to explore new approaches (these ideas have been adopted more recently by Boserüp, 1965, and Tiffen, 1993). Those who undertake ecological studies to explain human fortunes often place an overemphasis on adaptation and ignore the full spectrum of behavioural and other socio-cultural factors.

Biological or genetic determinism seeks to explain social and psycholog-ical developments as controlled by genetics rather than nurture. Drury

(1999: 3) argued that evolution is neither determined by genes, nor is it the outcome of pure chance, rather it is the interplay between the two. One difference between environmental determinism and new-environmental determinism is that the former mainly focused on occasional severe events like droughts, whereas the latter also considers everyday 'gentle' environmental controls that may gradually shape history in subtle ways. There are some parallels with punctuated equilibrium (familiar to those interested in biological evolution): long spells of gentle conditions result in gradual adaptation, and occasional sudden and severe catastrophes (for which normal evolution may have established little protection) may cause marked evolutionary 'jumps'.

An example of a new-environmental determinist is Couper-Johnston (2000: 143), who linked human affairs – including revolutions, social change, fortunes in war and even the development of the slave trade – with El Niño events: 'It is no coincidence that ... drought years often correlated with those years in which the greatest number of slaves were traded.' He also argued that the Polynesian settlement of the Pacific was made possible when prevailing **winds** briefly ceased in El Niño years, allowing their boats to access new areas, having been prompted to move by drought and famine also caused by these occasional ocean-atmospheric events (2000: 151–4). While such causation is possible, it is not adequately proven. Eisenberg (1998: 45) noted:

> ... every landscape in which humans have lived has sent up its characteristic schools of song ... every novel, poem, painting, and play has roots, at however many removes, in a landscape (or in several). One need not swallow the whole doctrine of environmental determinism to see that it has a kernel of truth.

Long before modern developmental psychologists began to argue over 'nature versus nurture' (i.e. genetic and biological versus environmental and social determinism), Thomas Hobbes (1588–1679) was exploring the issue, and almost a century later Jean-Jacques Rousseau (1712–78).

In the 1950s a group of intellectuals moved away from environmental determinism to the other extreme; these 'behaviouralists', like the psychologist B.F. Skinner, argued that an infant was effectively a 'clean slate' and learnt behaviour – so nurture (socialization) was crucial, far more so than environment (Skinner, 1972). Studies of human twins separated at or soon after birth have suggested that genetics are not always dominant.

Palaeo-anthropologists and archaeologists take a keen interest in how humans in the past may have related to their environment to explain physiological, and even cultural and mental development. During the 1970s Elaine Morgan generated interest (and some scepticism) with an explanation for human physiology that drew upon environmental determinism: an 'aquatic ape' hypothesis. Morgan (1972) developed ideas originally put forward by Sir Alister Hardy, that features – such as relative lack of body hair, upright posture, fat layers, structure of the windpipe, etc. – reflect an aquatic past. E.O. Wilson (1975: 145) observed that, at least with lower organisms, 'The most drastic response to fluctuations in the environment short of genetic change itself is modification of body form' (i.e. morphogenetic change). For example, young of particular insects may develop to fit a particular caste (which may mean a distinct morphology, as in ants) as a result of triggers outside the individual, and the sex of fish and reptiles may be determined by the temperature of the environment when eggs are incubating. Looking at the development and modification of social behaviour, Wilson recognized a range of responses, and noted that individual organisms track short-term environmental changes with physiological and behavioural responses, while the species they belong to must undergo evolution to cope with longer-term changes.

Supporters of environmental determinism today argue that there is a constant, complex and active interaction between living organisms and their environment, which includes other organisms. So social structures and their processes impinge upon the environment, modifying and changing it, and such changes in turn shape the structure of social systems, which subsequently undergo further changes (Norgaard, 1984). Social and economic responses to environmental change may be intended or unintended (or at least partly intended). E.O. Wilson noted that when a few individuals invent something that enhances survival and reproduction others may imitate and benefit (Wilson, 1975: 156).

The question of how far culture is inherited in the genes as opposed to learnt from others or prompted by environment is still a 'hot' issue. In the USA human behaviour genetics has run into fierce disagreements when there has been debate about genetic causes, rather than socio-economic oppression as the root of group differences in crime, etc. Some behaviour is clearly inherited (e.g. some schizophrenia and certain types of chronic depression). Gender determinism has been proposed, explaining that women inherit qualities that make them better at tasks like negotiation. There are linguistic determinists who claim language plays a major role in shaping thought and actions. Not surprisingly, the question is increasingly asked (technological determinism): 'How far does technology, including the media, determine social change?'

New-environmental determinism has been approached in various ways, and to various degrees; one being through 'bioregionalism' (Sale 1991; Frenkel, 1994; Mannion, 1994). This is an 'eco-movement' that first appeared in the 1970s and professes a commitment to the development of human communities in equilibrium with their ecosystems. Supporters of bioregionalism feel that humans are 'out of touch' with their environment and should 'return to living in harmony with nature' by adopting appropriate livelihoods and politics. Those adopting a Gaian viewpoint see humans as markedly constrained, if not severely limited, by the environment.

Post-war neo-determinists include Sauer (1963), Eyre (1978), Diamond (1997), Landes (1998) and Slack (1999). The latter three claim that Europeans have fared well largely because of environmental advantages. Diamond has been at pains to exclude race as a determinant, arguing that the marked differences between living standards in various parts of the world are not due to innate differences between peoples but to variations in the opportunities offered them by their environment. (For a critique of Diamond see http://aghistory.ucdavis.edu/Blaut.html, accessed 25 May 2000; Raudzens, 2001.)

Environmental historians argue that history is often shaped by natural events (Worster, 1988; Ponting, 1991). An increased interest in environmental history seems to have been prompted by growing bodies of evidence provided by palaeoecologists and ecologists. Also, the El Niño/ENSO (El Niño Southern Oscillation) events of the late 1990s have also helped renew interest in the environment–human relationship (Fagan, 1999; Couper-Johnston, 2000; Caviedes, 2001). Some have made sensational claims such as, 'For the last five thousand years, ENSO and the NAO [North Atlantic Oscillation] have killed millions of people, caused civilizations to collapse, and taxed human ingenuity to the limit' (Fagan, 1999: 90). However, neo-determinists more often tend to temper their claims by noting that environmental factors are 'possibly' the 'trigger' for a perhaps already ailing civilization, or whatever. Transhumance, fisherfolk and shifting agriculture (bush-fallow cultivation) are fields in which links between human movements and environment have been well explored.

Popular interest in neo-determinism has been promoted recently by authors like Fagan (1999) and Keys (1999). The latter has tried to 'reinstate respect for the basic concept of determinism – though not for the often simplistic nature of much past deterministic thinking' (1999: vii). Keys presents some interesting arguments for climate-induced geopolitical change, which he believes was ultimately caused by the eruption of a volcano near Java. Keys is not saying anything very new, Huntington (1915; 1919; 1945) and Toynbee (1976) tried to trace the rise and fall of civilizations and the migrations of peoples in response to climate, stressing the synchrony of often distant historical events – ergo a common environmental cause; unfortunately **dating** is often not precise enough to make good correlations and, as already stressed, even when there is a correlation it does not necessarily denote causation.

How useful, one might ask, is environmental history? It could certainly help us understand how the past unfolded, and this may give useful warnings and suggest strategies for the future. But, as Gould (2000: 211) warned, there are so many factors involved in human fortunes that attempts to make future predictions will be highly inaccurate. However, the past shows that humans are offered challenges and opportunities and prosper if they react quickly and correctly. Knowing that various threats and pitfalls exist should help administrators and those developing technology, infrastructure and governance to do things that help humans 'grasp the moment' and regain adequate adaptability.

Interpretation of environmental history must be cautious. Even though a repeated pattern of adverse weather, famine and plague recurs between AD 540 and the 1740s (perhaps correlated with a volcanic eruptions) it is not sufficient proof of direct environment–human welfare linkages; the course of history is influenced by a large number of variables that separate 'supposed cause from presumed effect' (Slack, 1999: 2).

Historical climatologists focus down on the components of climate to try and establish what influences humans, reviewing records for average temperature, extreme temperature, humidity, variability, barometric pressure, and so on (Brooks, 1942; Budyko, 1974). During the past 50 years or so, there has been a vast improvement in the data generated by both the physical sciences and environmental historians. This is partly the reason for renewed interest in the environment–human relationship (Pfister *et al.*, 1999). Also, the appearance of better computers and steadily improving monitoring, modelling and forecasting techniques should help those seeking to better map chains of causation (Moran, 1984; 1990). With expanding knowledge it is possible to trace the details of human–environment interaction, and to explore them with more reliability than ever before. Ellen (1982: xiii) argued that the way forward was through a systems approach with the 'focus on the integration of particular ecological variables, rather than the earlier attempts to obtain crude broad correlation between social organizations and environment'. Burroughs (1997: 133–4) was less optimistic, arguing that accuracy and speed of measurement and calculation does not help if there are very complex random events with even minor uncertainties. Others have also spotted that complexity, plus an element of randomness, is likely to thwart deterministic ambition – and coined the expression 'butterfly effect'. The reality is that even trying to model and understand just one simple element of the environment–human relationship can involve unmanageably vast quantities of data and might at best yield uncertain results.

One can argue that where there are strong environmental factors at work, as in harsh polar or arid desert environments, then physical conditions are more likely to dominate culture because there is a limit to what choice organizational skill and technology can give. Conversely, in less harsh environments, factors like culture may be relatively more important. Vandderbloemen (2000) adopts this sort of view, with environmental determinism and possibilism overlapping. With many now supporting the idea that the biosphere and human society are co-dependent there is some support for a sort of holistic environmental determinism. Fagan (2000: xiv) wondered to what extent the climatic shifts of the Little Ice Age (see Chapter 3) altered the course of European (and perhaps other peoples') fortunes, noting: 'while environmental determinism may be intellectually bankrupt, ... climate change is the ignored player on the historical stage'. Caution is needed to ensure that the enthusiasm to recognize climatic determinism does not miss other factors at play; some determinists may need to be reminded.

Human decision-making can be a very complex sequence of events, making it difficult to establish if there are environmental factors at work and, if so, how important these are. Le Roy Ladurie (1972: 17) recognized this and stressed that researchers must look carefully at the subtle ways in which climate change affects human activities (the same advice should apply to other environmental parameters). Brooks (1926) warned that there are many ways of interpreting environment–human relations, and that in any given situation more than one may seem plausible; writing two decades later, he noted that practically every aspect of human life is directly affected by climate (Brooks, 1950: 11). Delano Smith and Parry (1981: 3) stressed that space–time coincidence of climate change and socio-economic developments may not indicate causation – for example, if historical records show Alpine glaciers shrinking and settlement moving higher, is the human expansion caused by the ice retreat, or might it be due to population increase, civil unrest, land degradation or a combination of factors? Those studying environment–human interrelations must seek corroboration by as many different lines of evidence as possible before drawing conclusions. (An in-depth exploration of human choice within societal institutions in response to climate change was provided by Rayner and Malone (1998).) It is also important to realize that there have always been, and will continue to be, occasional random extreme climatic events that will defy forecasting.

There are pitfalls to trap those seeking to understand environment-induced human changes in the past; some natural cycles may behave in a reasonably predictable way for a long time and then markedly shift their pattern; it is also possible that a number of cyclic phenomena will interact occasionally through chance. Some environmental changes are so gradual and subtle that they are overlooked, even if they are important to humans. For example, the soils left at higher latitudes after glaciers retreated around 10,000 BP may have been slowly leached of some of their nutrients with few land users realizing; wind-blown dust may be deposited over centuries causing fertility changes; lakes undergo a successional development, perhaps almost imperceptibly altering from nutrient-poor and acidic to nutrient-rich and alkaline. Too much interpretation is still speculative. Often the best assessment is to offer a number of more or less probable explanations (Goudie, 1977: 202).

Crude environmental determinism has been discredited. However, there is a need for caution not to 'throw the baby out with the bath water'; there may be cases in the past where human fortunes have been *affected* by environmental changes, even if they were not *determined* by it. Environmental deterioration tests society – people may adapt well, barely muddle through, escape by migration, or fail and suffer. In the past humans often relied on moving, and on strong social institutions and social capital to adapt; for the last two centuries the developed countries have depended more on technological innovation and centralized government, and have to a large degree lost social capital and the ability to move.

Key reading

Barry, J. (1999) *Environment and Social Theory*. Routledge, London.

Diamond, J. (1997) *Guns, Germs and Steel: a short history of everybody for the last 13,000 years*. Jonathan Cape, London.

Glacken, C.J. (1967) *Traces on the Rhodian Shore: nature and culture in western thought from ancient times to the end of the eighteenth century*. University of California Press, Berkeley (CA).

Le Roy Ladurie, E. (1972) *Times of Feast, Times of Famine: a history of climate since the year 1000* (trans. from the French by B. Brag). George Allen & Unwin, London (1971 edn, Doubleday, New York NY; originally published in French, 1967, as *Histoire du climat depuis l'an mil*. Flammarion, Paris).

McNeill, J.R. (2000) *Something New Under the Sun: an environmental history of the twentieth century*. Penguin, London.

|2|

Environmental change before modern humans appeared

The key to the future is the past.

(Officer and Page, 1998: vi)

Chapter summary

This chapter reviews environmental change that took place before modern humans appeared. Historical records, aural traditions and myths give some indication of environmental challenges and impacts; however, there may be those that have not happened within memory, but that recur over a longer timespan – looking far back can thus be invaluable for understanding the present human and environmental condition, and for aiding future scenario prediction. There have clearly been a number of major extinction events, huge volcanic eruptions and the catastrophic collision of extraterrestrial bodies with the Earth; and, during the last few million years, considerable climatic and sea-level variation.

Concentrating only on environmental changes since the evolution of humans would be to miss events that, although they happened millions of years ago, might recur in the future. Understanding the past is also an aid to comprehending present processes and predicting outcomes. For example, meteorological records were not enough to clarify the causes of the damaging Sahelian drought of the late 1960s and early 1970s; researchers also explored palaeoecological evidence (Brooks, 1926; 1950; Lamb, 1988; Calder, 1974; Gribbin, 1976a; 1978; Ingram et al., 1981). This chapter focuses on events before roughly 5000 years BP.

Written records perhaps date as far back as 4000 BP; myths and legends may provide some hints about the past, perhaps to around 12,000 BP; but,

Table 2.1 The geological timescale

Eon	Era	Period	Epoch	Age	million years BP
P H A N E R O Z O I C	CENOZOIC	Quaternary	Holocene (Recent)	0.01	
			Pleistocene	1.64	
		Tertiary ■	Pliocene	5.2	
			Miocene	23.3	
			Oligocene	35.4	
			Eocene	56.5	
			Palaeocene	65.0	KT-boundary
	MESOZOIC	Cretaceous	Senonian		
			Gallic		
			Neocomian	144.0	
		Jurassic	Malm		
			Dogger		
			Lias		
		Triassic	Late Triass		
			Middle Triass		
			Synthian	248.0	
	PALAEOZOIC	Permian	Zechstein		
			Rotliegendes	295.0	
		Carboniferous*	Stephanian		
			Westphalian		
			Namurian		
			Visean		
			Tournaisian	354.0	
		Devonian	Late Devonian		
			Middle Devonian		
			Early Devonian		
		Silurian	Pridoli	416.0	
			Ludlow		
			Wenlock		
			Llandovery	442.0	
		Orodovician	Ashgill		
			Caradoc		
			Llandeilo		
			Llanvirn		
			Arenig		
			Tremadoc	495.0	
		Cambrian	Merioneth		
			St David's		
			Caefai	544.0	
PROTEROZOIC (CRYPTOZOIC)				2500.0	
ARCHAEAN				3800.0	
PRISCOAN (HADEAN)				4650.0	

Notes
BP (before present) = before AD 1950
Tertiary-Quaternary boundary is officially placed at 1.8 million years BP.
■ = divided into Neogene and Palaeogene (division at 23.3 million years BP). The Tertiary is no longer formally recognized, but remains in widespread use.
• = divided into Pennsylvanian and Mississippian (widely used outside Europe).
Life had probably appeared by 3,800 million years BP. Atmospheric oxygen levels probably reached similar levels to the present day roughly 500 million BP.
Proterozoic and Archaean sometimes grouped as pre-Cambrian.
Archaean/Hadean-boundary is difficult to date because erosion and metamorphism has destroyed datable rocks.
Terrestrial vertebrates appear after 408 million BP.
Source: Author.

for most of the time humans have been evolving, there is no memory, and if there were it would still be too short to cover those rare but important Earth events that have not recurred since we appeared. Hominids were evolving by 6 million years BP and creatures similar to modern humans date back roughly 2.5 million years or more – a small slice of time, given that the Earth may be over 4.6 billion years old (see Table 2.1). It is important to look back to the past before humans appeared in order to get an idea of possible future challenges and to explain the responses and adaptations organisms have already made.

When the Earth newly formed (roughly 4600 million years BP) it is assumed that there was an atmosphere rich in hydrogen, methane, ammonia and carbon dioxide, possibly as much as 60 times as dense as today. Single-celled life forms had probably appeared by 3870 million years ago and photosynthetic bacteria and algae altered the primordial atmosphere to one richer in free oxygen similar to today's, somewhere between 2500 and 650 million years BP. Multicellular organisms had evolved before 540 million years BP (that point in time being the so called 'Cambrian explosion', a point at which a diversity of life forms, relatively suddenly, developed qualities that improved their chance of being fossilized). The large flying insects and huge sea scorpions of the Devonian and Carboniferous (roughly 350–290 million years BP) strongly suggest higher oxygen levels than now (probably about 35 per cent compared to the present-day 21 per cent). Later, after vegetation spread on to dry land, these oxygen levels must have fallen, or there would have been vast bushfires; 'red beds' (iron-rich sedimentary rocks) were being deposited and have been interpreted as evidence that oxygen was being sequestered from the atmosphere, probably by bacteria (Drury, 1999: 137). Oxygen levels less than 10 per cent to 16 per cent would make it difficult for higher life forms to function, and levels above 22 per cent approach the point at which severe bushfires become a constant threat. Oxygen concentrations seem to have remained relatively stable around 21 per cent for roughly the last 300 million years.

Atmospheric carbon dioxide and methane levels seem to correlate with global climate. The record of past carbon dioxide and methane changes, together with a range of other geochemical, geological and palaeontological indicators, can be used to deduce past climate. Ignoring the very slow changes caused by the gradual movement of landmasses (continental drift), there appears to have been marked climatic change, some of which may have been periodic. Earth history suggests that for perhaps the last 600 million years conditions have usually been significantly warmer than now. The warmth has been punctuated by a number of glacial epochs, some mega-ice age 'icehouse' conditions, and others less severe ice ages (Imbrie and Imbrie, 1979). Bryant (1997: 13) suggested that these glacials each lasted roughly 700,000 years and each can be subdivided into approximately 96,000-year-long cold phases separated by warm interglacials. These glacial epochs, or ice ages, may recur every 250 to 308 million years or so, possibly reflecting the rotation of our galaxy (Gribbin, 1976b), convection in the Earth's mantle, plate tectonics, periodic shifts in oceanic mixing or changes in world sea levels (see Box 2.1). Clearly, the Earth does not need human pollution to trigger considerable climatic variation.

Box 2.1 Icehouse and glacials

2.4 to 2.3 billion years BP (pre-Cambrian) – some signs there was (icehouse) glaciation to within 10 degrees of the Equator; perhaps total ice cover. Opponents of total (worldwide) icehouse conditions argue that ocean circulation would make it difficult. At present, geological evidence is limited and climatologists and biologists are sceptical. The Gowgonda glaciation is placed at 2288 million years BP (+/– 87 million years).

900 to 610 million years BP (late pre-Cambrian) – good evidence of glaciation on the palaeocontinent of Gondwanaland. The Infracambrian II glaciation is placed at 950 million +/– 50 million years BP, and the Infracambria I at 740–750 million +/– 40 million years BP. The Eocambrian glaciation is placed at 650 million +/– 50 million years BP.

750 to 600 million years BP (Permian) – (icehouse conditions) ice cover may have reached the tropics, and may even have been worldwide (the latter is fiercely debated, not least because such conditions would have made the survival of more complex life forms unlikely). There seems to have been a number of glacials and interglacials over, perhaps, 200 million years. While there may have been extensive ice cover in the Neoproterozoic (around 700 million years BP), the Equatorial oceans probably failed to freeze over (*Geology* 29: 1135).

440 million years BP (late Ordovician) – glaciation may have extended into the tropics (near-icehouse); possibly almost worldwide. The evidence is for repeated glacials taking place during a c. 50 million-year period. The Siluro-Ordovician glaciation has been placed around 410–470 million years BP (c. 453 to 438 million BP according to some sources).

350? million years BP (late Devonian) – some evidence of glacial conditions from Brazil.

300 million years BP (Carboniferous) – possibly ice sheets in the Southern Hemisphere. The Permo-Carboniferous glaciation has been placed at 255–340 million years BP.

280? to 270? million years BP (late Carboniferous to mid-Permian) – repeated glacials.

55 million BP (Oligocene) – ice starts spreading at poles.

22 million years BP, perhaps a little earlier (early Miocene) – there are signs of global cooling by roughly 10 million BP, which seems to have gradually led to the mid- and high-latitude glacier and ice cap formation in the Quaternary era. There was mountain glacier formation even in the tropics by 3.5 million BP. Ice sheets had probably started to grow in high-altitude areas of Antarctica by about 14 million BP. At the peak chill of the last Quaternary ice age (roughly 33,000 years BP) average global temperatures were probably 7°C to 13°C below those of the present. The period 0–14 million years BP is sometimes referred to as the Cainozoic glacial.

Source: Compiled by author from several sources.
Notes: A billion is one-thousand million; dates are very approximate; BP = before present (1950); the start of the Quaternary is officially placed at 1.8 million years BP (on stratigraphic criteria).

Humans have largely evolved during the most recent ice epoch: the Quaternary Era. This began (depending on how one dates it) between 3.0 and 1.8 million years BP; and continued to roughly 10,000 BP. According to estimates by Bryson and Murray (1977: 127), permanent ice covered about 9 per cent of the Earth's surface at the last glacial maximum (roughly 33,000 years BP), compared with about 3 per cent coverage today. There is still no firm agreement on how many advances and retreats of ice occurred during the Quaternary – there may have been 50 or more cold stages, some separated by warm (interglacial) stages; the former were of about 40,000 years' duration, the latter each spanned about 20,000 years, although some were as long as 100,000 years (John, 1979; Dawson, 1992; Crowley and North, 1991; Mannion, 1999: 29–31). During the glacial stages global average temperatures fell roughly 9°C below the present. The glacial stages were sometimes punctuated by warmer periods (but not full interglacials) – interstadials; the warm stages were broken up by cold periods (but not full glacials) – stadials. This Quaternary ice epoch was tame compared with some in the more remote past; there is evidence that at least one pre-Quaternary icehouse period (during the Permian) was so severe that sea ice reached the tropics and perhaps wholly covered the Earth, causing a mass extinction. Another similarly extensive glacial probably occurred earlier at the end of the Ordovician, *c.* 440 million years BP.

It has been suggested that the Earth can adopt one of three equilibrium modes (and it would be interesting to see if Mars and other planets have had comparable fluctuations on a similar timescale):

ice world – ('icehouse') the cold extreme with ice cover over most of or all of the globe
glacial–interglacial fluctuations – polar, high-latitude and mountain ice cover roughly 10 per cent more extensive than now, with depressed global temperatures (roughly 6°C to 10°C below the present); the world is probably in an interglacial phase of this mode today
ice-free world – ('greenhouse') the warm extreme; due possibly to more carbon dioxide in the atmosphere than today; greenhouse conditions in the Cretaceous resulted in global mean temperatures up to 15°C above present.

Current fears about anthropogenic greenhouse warming are still justified because all but the most gradual change can disrupt livelihoods and infrastructure in a catastrophic manner; also, rising temperatures could trigger runaway warming to a life-threatening level, or might prompt a stadial (cooler period). Palaeoecology shows that life can flourish under considerably higher global temperatures than now; however, organisms (including humans) need a lot of time to adapt. Anthropogenic warming that caused an ice-free world (mode 3 above) would probably be far preferable to a natural swing back to glacial conditions of mode 2. Although a mode 2 glacial would expose considerable areas of land as sea-levels fell, it is doubtful that modern humans could adapt fast enough to benefit before environmental changes caused disaster. Gentle greenhouse warming, like that taking place at the moment, may cause economic and other miseries for centuries or even initiate a shift of the Gulf Stream and, for a while at least, badly chill western Europe.

Warm 'greenhouse' conditions (as much as 15°C above present average global temperature) prevailed for at least 100 million years before the late Tertiary, when the cooling that culminated in the Quaternary ice age started. In the Cretaceous (before *c.* 65 million years BP) coal deposition was taking place close to the North Pole, well beyond present vegetation limits, and lowland Antarctica was forested until 16 million BP (perhaps retaining vegetation until as late as about 6 million BP). Ice caps had started to form in Antarctic highlands between 25 to 30 million years ago. The last appearance of subtropical palm trees in Europe north of the Alps (in the Lake Constance region) was in the late Miocene (*c.* 2.6 BP), when ocean temperatures were falling and, around the world, mountain glaciers and polar ice caps were expanding.

Modern humans should be aware that there have been repeated 'mass extinction events' and that there is a risk of them in the future; the most serious took place in the late Devonian, at the Ordovician–Silurian boundary, and at the Cretaceous–Tertiary boundary (the KT-boundary) (see Box 2.2). Mass (or major) extinction events are relatively sudden catastrophes that allow little time for biological adaption (organisms can adapt a great deal *if* change is not too fast). That these mass extinctions have happened is widely accepted, but the causes are uncertain.

Box 2.2 Mass extinction (major extinction) events

Researchers have so far recognized up to twelve, but more clearly eight, mass extinction events.

Million years BP approx.

- *2,500–2,000 Proterozoic era* – photosynthetic bacteria and algae generate oxygen. Oxygen-sensitive organisms die out or retreat to extreme, oxygen-free environments.
- *600 Mid-Cambrian* – marked loss of species. Some suggest that the Earth became completely ice covered from poles to tropics for around 10 million years before thawing through carbon dioxide accumulation in the atmosphere. If this did happen, primitive life forms, such as algae, must have survived in open water near the Equator or beneath the ice sheets.
- *438–440 Late Ordovician* – possibly another worldwide or nearly worldwide freeze.
- *350 Late Devonian (Frasnian-Famennian event)* – this almost killed off a highly successful and adaptive group of marine arthropods (now extinct), the trilobites. Asteroid strike or cooling (oceanic carbon dioxide records suggest the latter)?

- *246–250 Late Permian (Guadaloupian event)* – possibly the most severe extinction. It may have wiped out over 95 per cent of the Earth's life forms (more severe than the KT-boundary event). Coincided with huge sheet-lava outpourings in Siberia. This event marked the end of the trilobites, and coincides with a carbon-12 to carbon-13 anomaly in marine and terrestrial sediments (perhaps indicative of severe environmental change). Volcanic activity might have over-heated or, more likely, chilled the Earth.
- *208 Late Triassic (Norian event)* – perhaps 48 per cent of marine genera were lost.
- *66.5–70 Cretaceous-Tertiary (KT-boundary) event* – marks the end of the dinosaurs. There is speculation about the cause being an asteroid or comet impact. However, a severe volcanic event, perhaps the eruption of the Deccan Traps of India or other causes cannot be ruled out.
- *34 Eocene-Oligocene* – a widespread iridium layer, possibly indicating an asteroid strike.
- *14 Mid-Miocene* – may be related to the Ries Impact Crater in Germany? Not well documented.
- *5 million BP* – extinction event? (Not a full mass extinction.)
- *2 million BP* – extinction event? (Not a full mass extinction.)
- *Present day (post-AD 1700)* – some ecologists argue that the Earth is undergoing a mass extinction, with human activity the cause.

Notes

1 Before the Cambrian there were probably extinction events but erosion, subduction and metamorphism have destroyed the evidence; also, organisms were mainly small and soft-bodied and left little fossil record. Before roughly 350 million years BP life was confined almost entirely to the seas.

2 Some researchers argue that at least some of the extinction events took place in phases or steps over quite an extended period.

3 Permian-Triassic (PT-boundary) event possibly the second most severe extinction since the Permian.

4 Two Cretaceous mass extinctions have been recognized: the KT-boundary event and an earlier (Maastrichian) event, which was marked by the loss of marine organisms including the ammonites and belemnites.

5 The Deccan outpourings occurred at about the right time and lasted long enough to explain the relatively slow extinction of organisms. Recently the rare metal iridium has been found in material ejected by a few volcanic eruptions, so it may not only be indicative of extraterrestrial body strikes.

6 These extinctions might be periodic: occurring roughly every 26 to 36 million years. Drury (1999: 292) noted the same 'periodicity' for mass extinction events and flood basalts over the last 300 million years; this he felt is statistically unlikely to be random. There are also signs that oxygen isotope records show a similar periodicity over the last 130 million years. It has been suggested that this reflects the recurrent passage of the solar system through the plane of the Milky Way galaxy every 26 to 32 million years during each c. 250 million-year 'Galactic year'.

Source: Compiled by the author.

2.1 The need to understand possible causes of past extinctions

If the cause (or causes) of past mass extinctions can be established, humans might be able to prepare for, and avoid or mitigate, similar future events. A number of possible causes have been proposed, none are proven. Some may be so gradual that they could easily be overlooked; it is even suggested that humans themselves are responsible for an ongoing mass extinction today – there has certainly been great loss of biodiversity in the past few centuries due to land development, hunting, pollution and so on. The possible causes of past extinctions might be abiotic or biotic, endogenous or exogenous (terrestrial or extraterrestrial), relatively slow and gradual or sudden and swift, single events or multiple events. Mass extinctions might result from one or a combination of causes; they might show some pattern, or may have occurred in a more or less random manner. Causes might include one or more of the factors described below.

Serious volcanic eruptions

These are either relatively non-explosive but very extensive outpourings of flood lavas (plateau basalts or traps), or violently explosive. Some flood lavas are hundreds of metres thick and cover thousands of km². Violent eruptions can project huge amounts of fine ash, CO^2, sulphate-rich aerosols and gases like sulphur dioxide into the atmosphere. Both types of vulcanicity may eject enough acid aerosol compounds and gases to form a 'dust-veil' that cools the world (something Benjamin Franklin in the USA and also Gilbert White in the UK suggested briefly happened after the 1783 Laki eruption in Iceland), may cause widespread acid deposition and perhaps even make the atmosphere difficult to breathe. How devastating the effects are and how long they last depends on the characteristics of the eruption or eruptions. In modern times some climatic cooling has followed certain eruptions for a few years but in prehistoric times huge eruptions may have caused chills lasting thousands of years – possibly even icehouse epochs.

Outgassed CO_2 raises global temperatures if it is not 'locked up' by growing plants or deposited as carbonate rocks; it is possible that some large eruptions caused greenhouse conditions in this way, rather than cooling. Examples of vast flood lavas include: the Deccan Traps, India (*c.* 66 million years BP), which are extensive basalt lavas covering 2.5 million km²; the Columbia Plateau Lavas, USA (*c.* 247 million years BP); the Columbia River Basalt Group – Roza Flow (USA, *c.* 14 million years BP), with an estimated volume of 700 km³.

These mega-eruptions and outpourings call into question a wholly uniformitarian worldview; at times the planet has been much more volcanically active than it is now. Courtillot (2000) argued that at least seven mass extinction events coincide with intense volcanic activity, and he is not alone in making this observation (Officer and Page, 1998). It has been suggested that the KT-boundary event was due to mega-eruptions, rather than the other most favoured cause: asteroid or comet strike. The KT-boundary event might have happened when a plume of hot magma penetrated up through the Earth's mantle, causing huge fissure eruptions that formed the Deccan Traps. In the eighteenth century AD a vastly smaller fissure eruption in Iceland ejected toxic gas and aerosols (especially fluorine) killing most of the island's livestock and blighting agriculture as far away as Scandinavia, Scotland and possibly western Europe, and it may have chilled the world enough to have caused a few years of famines. Recent, relatively modest, eruptions like Mount St Helens (USA, 1981) and Mount Pinatubo (the Philippines, 1991) have been monitored and shown to cause significant cooling of the world's climate for seven years or more. Past eruptions could have had greater and longer impact.

Whether there is any periodicity in the shorter term or the long term for volcanic activity has not been established; there may be some relationship between eruptions and the position of the Earth in relation to the Moon, various planets or the Sun. Alternatively, it might relate in some way to sea-level variations or the stress caused by ice build-up and deglaciation during glacials.

Evolution of organisms

Albritton (1989: 162) claimed that three mass extinctions correlate with evolutionary developments in land plants:

1. the spread of vascular plants on land in the Devonian
2. the luxuriant growth of pteridophytes (ferns) and gymnosperms in the late Carboniferous
3. the rise of the angiosperms in the Cretaceous (vascular plants are those with a system of vessels to conduct fluids; the gymnosperms include the conifers; angiosperms are 'flowering' plants, and include the grasses).

Each of these developments may have upset the Earth's carbon balance, causing climatic change and possibly altering the atmospheric gas mix (which would affect metabolism). They might also have locked up nutrients and consequently have stressed plankton in the sea, disrupting the food chain, which would starve higher life forms and trigger fluctuations in atmospheric carbon dioxide levels.

Sea-level changes

Sea levels change, perhaps due to variations in the shape of the ocean basins, or through climatic shifts. Changing sea-levels would have altered marine environments and the way terrestrial climates were moderated. It has been suggested that bodies of water like the Mediterranean could have been cut off, evaporated and enough salt taken out of the rest of the world's oceans to affect salinity and thereby **currents** and the freezing point at the poles, which in turn led to climate changes. This appears to have happened to the Mediterranean around 5.8 million years BP.

Continental drift

Although slow, this may link or separate landmasses or oceans, allowing altered species competition, triggering climatic changes and affecting the location and type of volcanic eruptions.

The effects of orbital variations

Changes in the Earth's orientation, or 'wobble', over time may alter the climate to cause extinctions. One theory that now has considerable support is that of **Milankovich**, which argues that shifts in the Earth's eccentricity, obliquity and procession 'drive' glacial–interglacial changes *within* an ice epoch. The Milankovich theory explains, a 22,000-year, a 40,000-year and a 100,000-year climatic periodicity (Pearson, 1978: 20–5; Bryant, 1997: 10).

Mountain uplift (orogenic activity) and other tectonic changes

The formation of ranges like the Himalayas means greater interception of rainfall, which takes carbon dioxide from the atmosphere and sequesters it in the seas as carbonate deposits. With high mountain ranges this sequestration may be enough to start an ice age. Mountains also alter climate downwind, forming a rain shadow that might have further climatic impacts. There have been a number of periods of greater than normal mountain building and volcanic activity.

Impacts on Earth by extraterrestrial bodies (planetesimals or near-Earth objects (NEOs))

Collision with sizeable rocky or metallic asteroids, and icy or carbonaceous comets could result in an air blast sufficient to devastate large areas; the impact might also splash fragments of molten rock, which would set off fires on a continental or even global scale; vaporized rock and water are likely to cause shading, leading to a worldwide 'nuclear winter' (global cooling for weeks to several months). These disturbances might also disrupt the stratospheric ozone layer resulting in increased irradiation of plants and animals, and could cause acid deposition, methane outgassing or huge tsunamis (if the strike is at sea).

Such strikes remain a significant threat. But, for the first time in the Earth's 4600 million-year history, humans have the potential to reduce the risk through technology.

Geomagnetic variation

It has been known that the Earth's magnetic field varies since the work of Gauss in the 1830s. Variation in the Earth's geomagnetism includes alterations in strength, field pattern (manifest as the 'wandering' of the position of the magnetic North Pole, which can be useful for approximately dating events during the Quaternary) and reversals of polarity (i.e. the magnetic pole moves from one hemisphere to the other). Reversals can be traced back hundreds of millions of years; however, the pattern becomes difficult to decipher accurately before about 5 million BP (Fig. 2.1). The reversals take place about three times in a million years and each takes about 5000 years to complete. Between each polarity reversal there seem to be around 20 fluctuations in intensity (Pearson, 1978: 65).

It is important to study these reversals to see whether there are signs of any past environmental changes associated with them that might recur in the future. Although many climatic changes in the past appear not to have been associated with geomagnetic reversals, there do seem to be links between increasing geomagnetic field and cooling. Possibly during geomagnetic variation there might be periods when there is increased penetration of solar radiation. It has been suggested that geomagnetism varies with fluctuations in solar activity, correlating with the quasi-periodic 'cycle' of sunspot activity. The Sun's outbursts – increased solar wind or plasma – might interact with the Earth's magnetic field (it is still not clear whether exogenous or endogenous causes alter geomagnetism, although the latter is favoured).

Variations in geomagnetism might reduce the amount of stratospheric ozone, thereby weakening the Earth's shielding from solar radiation. There have been suggestions that, at lower latitudes, this could lead to something like a 14 per cent increase in UV radiation penetration to the surface, enough to have significant effects on vegetation, terrestrial wildlife and marine plankton. Apart from killing sensitive organisms, the resulting increase in mutation rates might help speed up species evolution. The irradiation would probably be greater at higher latitudes and in uplands. There are some indications that past mass extinction events might coincide with some magnetic reversals.

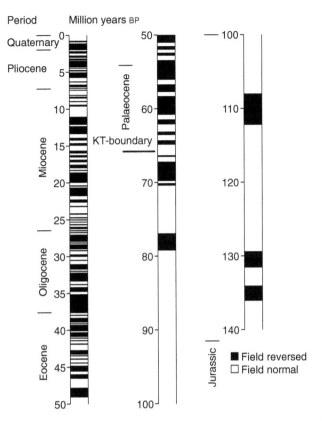

Figure 2.1 Changes in geomagnetic polarity back to the Jurassic (*c.* 140 million years BP)
Note: The last reversal to present-day polarity was about 730,000 BP; since the last reversal, the position of the north magnetic pole has wandered around the high latitudes.
Source: Reproduced and modified with permission from Pearson, R. (1978) *Climate and Evolution.* Academic Press, London and New York.

There are currently suggestions that a reversal event is due within 2000 years. Further research is needed to determine the effect of reversals and their cause. Once these are clarified mitigation measures might be possible (see Fig. 2.1).

Severe disruption of the Earth's stratospheric ozone layer

This would lead to increased receipts of UV radiation, and possibly climatic disruption. Causes could include certain volcanic eruptions, fluctuations in the Earth's geomagnetic field, unusual solar activity and gamma-ray bursts from collapsing stars. It is also possible that marine plankton might sometimes generate enough ozone-scavenging chemicals to thin the stratospheric ozone layer. Radiation would damage organisms on land but probably not penetrate deep enough into the sea to destroy all marine life. However, an extensive kill of plankton might explain the extinction of some organisms inhabiting deep and shallow water, like the trilobites, as it would have hit the food chain and their planktonic larval stage.

Human activity has been damaging stratospheric ozone for the past 70 years, mainly by the manufacture and release of chlorofluorocarbons (CFCs).

Altered Earth–ocean–atmosphere interaction

Variations in the flow of ocean currents caused by temperature or salinity changes might have marked effects on global climate. ENSO events (see Chapter 5) and altered thermo-haline circulation in the North Atlantic during the Quaternary have been attracting considerable research, because they may be the key to marked and sudden climate shifts. Changes in depth of water over continental shelves and the opening or closure of channels may also have strong climatic effects (the effect of isolation of seas like the Mediterranean have been mentioned above). For example, the slow closure of the Tethys Sea by continental drift during the Tertiary probably had significant effects on Eurasia. Alterations in cloudiness, vegetation and snow cover change the albedo (reflectivity) of areas and may affect climatic conditions. Vegetation changes might have marked climatic effects; in the past the appearance of new plants and grazing animals may have caused such changes, today forest clearance (notably in Amazonia and West Africa) may be affecting climate.

Large-scale death of marine plankton

This could have a number of causes: ozone depletion, water pollution through volcanicity or a strike by an extraterrestrial body (**planetesimal** or near-Earth object (**NEO**)), wind-blown dust from land, plankton species competition or an icehouse glacial. Loss of plankton would hit the marine food chain and perhaps reduce atmospheric oxygen. Plankton death is likely to be a symptom of some other global perturbation like increased incidence of solar radiation.

Increased receipts of cosmic radiation from outside the Solar System

This could possibly be due to a relatively 'close' stellar collapse (supernova), leading to a gamma-ray burst. Such an event would cause the irradiation of organisms, especially those on the Earth's land surface and in shallow marine environments (and would affect deeper-dwelling creatures if they feed in surface water or their young grow up there). A particularly strong or close event might even virtually sterilize the Earth, and destroy oceans and atmosphere. Astronomers and geologists think these events are likely to happen regularly, if infrequently, over geological time. Astronomers at the University of Texas estimate that probably around 1000 gamma-ray bursts intense enough to significantly affect organisms have struck the Earth since life began. These events are probably very brief, and so are likely to cause damage to organisms indirectly – for example, by thinning stratospheric ozone so that more UV radiation can penetrate the atmosphere. There is some evidence of such an event around 5 million BP and at approximately 2 million years BP (possibly as a result of a supernova in the Scorpius-Centaurus OB star cluster). One defence would be to colonize as many distant star systems as possible (beyond humankind's technological abilities).

Increased dust in space

This would cause a reduction of the relatively constant output of solar radiation reaching the Earth and might take place at regular intervals (*c.* every 250 to 308 million years) as the galaxy rotates, or at random if there are scattered dust clouds in the cosmos (there are signs of cyclic variation in micrometeorites in lunar soil-core samples, which may support the idea of galactic rotation).

Fluctuations of solar radiation

Increases (longer-term and more gentle, or brief but intense solar flares) or decreases in the solar 'constant' could have serious consequences for life on Earth. The cause might be the progressive consumption of elements or some other fluctuations within the Sun affecting its fusion (Waple, 1999). Sunspot abundance has long been known to show periodicity – there are good records for AD 1500 to the present, the interpretation of which has led to several suggested periodic cycles, including those of *c.* 22 years and a mean of *c.* 11.2 years (the Hale cycle).

There are several other recognizable periodicities, including *c.* 5.5, 10–12, 22–23, 80–90 and 170–200 years (Pearson, 1978: 44). What effect sunspot fluctuations have upon humans is uncertain, although some environmental historians have linked agricultural productivity and fish spawning to sunspot periodicity (Pearson, 1978: 49). Sunspots increase when the Sun is more active and may coincide with reduction of stratospheric ozone. Low sunspot activity appears to correlate with cold conditions on Earth, and high with more storms and lightning.

The output of solar radiation might vary because of factors beyond the Solar System – there has been a suggestion that the Sun might have a sister or companion star (the Nemesis hypothesis), and periodically their movements result in tidal or other effects that raise or reduce solar radiation (and perhaps increase the likelihood of extraterrestrial bombardment on Earth as the Sun's hypothetical companion shifts asteroids out of the Oort Cloud that surrounds the outer Solar System). This *might* explain the eight or so mass extinction events that are believed to have occurred since the Permian: roughly every 26 million years (Raup, 1986). If true, the Earth is probably relatively safe for roughly another 15 million years.

More research into solar radiation would be a useful investment. Already, telecommunications, power transmission (a 'solar storm' shutdown of the Canadian power grid and a blackout of New York State, USA, has happened in the past ten years), and electronic components in anything from vehicle ignition systems to satellites are vulnerable.

Increased deposition of iron compounds in the ocean

These are possibly derived from oxides blown from dryland areas, and could stimulate marine plankton and raise carbon dioxide fixation, leading to global cooling. There have been suggestions that humans could do this to counter global warming – a risky thing to attempt without better knowledge.

2.2 Environmental change between roughly 2 million years BP and roughly 5000 years BP

The last few million years of Earth history have been very variable, although modern development has been during a period of relatively benign and stable conditions; it is likely that climatic changes, sea-level fluctuations and the movement of game animals repeatedly separated and then brought together hominid populations, ensuring genetic drift and ongoing biological, cultural and technical evolution, ultimately leading to *Homo sapiens* (Pearson, 1978; Binford, 1999). Even with a growing body of evidence from palaeoecologists, palaeoanthropologists and archaeologists it is only possible to speculate about the environment–human relationship for much of the last few million years. Changing climate might have played a part in prompting hominids to leave forests for the savannahs and capitalize upon an upright, two-legged stance; although bipedalism could also have evolved as a result of wading through water or high grass, or because of behavioural reasons unrelated to environmental causes. The likelihood is that climatic changes will at the very least have prompted repeated movement from area to area, and changes in behavioural and survival strategies.

The evidence of past climatic change has vastly improved in the past few decades; the record now stretches back further, in far more detail, and with greater accuracy than could have been imagined even in the 1950s. Studies of environment–human relations have been pursued using ice cores, dendrochronology (tree-ring studies), ocean sediment records (see Table 2.1 and Figs 2.1 and 2.2), palaeofauna and palaeoflora studies, geomagnetism and much more. Some of the most detailed evidence for environmental

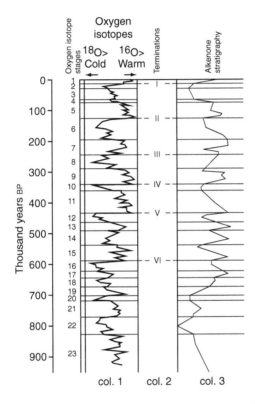

Figure 2.2 Ocean sediment record – core from eastern Equatorial Atlantic
Key
Column 1 presents the oxygen-18 : oxygen-16 record – calcareous-shelled plankton incorporate oxygen while living; the greater the amount of heavier oxygen-18 in the fossil shells, the colder the conditions at the time of deposition.
Column 2 shows the last six 'terminations' (I is the end of the last glacial; II the end of the penultimate glacial, and so on); these are sharp, sudden changes from cold to warm conditions.
Column 3 shows the alkenone unsaturation index. Alkenones are a component of the lipids (fats) in plankton, a high unsaturated : saturated ratio indicates a warm sea-surface temperature. Note the broad agreement of the curves in columns 1 and 3.
Source: Redrawn and modified with permission from Mannion, A.M. (1999)
Natural Environmental Change. Routledge, London.

change, back to 110,000 BP or earlier, has come from ice cores like GISP2 (US Greenland Ice Sheet Project 2), one of the most successful of the cores so far recovered from the Greenland ice cap, and GRIP (European Greenland Ice Core). Others have been bored from Antarctica and a number of mountain glaciers (see Mayewski, 'GISP2 and GRIP records prior to 110kyr BP' at http://www.agu.org/revgeophys/mayews01/node8.html, accessed 25 May 2000). The cores indicate that climate change has sometimes been very rapid, at least in the North Atlantic sector, where the Younger Dryas cool phase terminated abruptly over less than ten years with a rise in temperature of 7°C.

By the 1780s a number of researchers were becoming increasingly aware that there had been glacial conditions in the past. De Sassure recognized erratic boulders, a sign that glaciers in the Jura Alps had been more extensive in the past, and by the 1830s a number of scientists, including Agassiz, Playfair, Hutton, Venetz, de Charpentier and Brückner had established the idea that there had been a Quaternary ice age. It is now clear that much of the development of modern humans has been shaped by glacial episodes, other marked climate and sea-level changes, and frequently altering floras and faunas; it is fair to say that 'Man is child of the Quaternary' (Charlesworth, 1957; Fagan 2000: xii; Imbrie and Imbrie, 1979).

During the last *c.* 2.6 million years, temperatures have fluctuated between interglacials with the global mean as much as 4°C warmer than today, and glacials with the global mean as much as 9°C colder than today. In the tropics temperature falls corresponding to glacials in higher latitudes lowered vegetation zones in uplands and in some areas caused rainfall changes at lower altitudes. An idea of the scale of ice accumulation during glacials is given by the fact that the Laurentide Ice Sheet (North America) was probably over 12,000 feet (*c.* 4,000 metres) thick and the Fennoscandian Ice Sheet (northwestern Europe) was probably more than 9000 feet (*c.* 3000 metres) thick.

The last glacial 'peaked' around 18,000 to 20,000 years BP, and melting was well under way by 10,000 years ago. The shifts in ice cover, changes in sea-level, and altered positions of weather systems during the fluctuations between glacial and interglacials have had profound global effects on ocean circulation and terrestrial climate. The last interglacial (the Eemian, roughly 115,000–125,000 BP) appears at times to have been warmer than the present, although conditions were not especially stable and there were marked climate variations during it.

Culture and technology helped cushion the effects of Quaternary environmental change on humans. In particular, the development during the past 1.5 million years of clothes, flint weapons and tools, use of fire, boats, herding and agriculture (perhaps excepting the latter), gave more chance for adaptation and survival. Tools appeared during the Pliocene before the climate deteriorated, and were at first barely distinguishable from natural stones. Between 3.5 to 2.4 million years BP there is reasonable evidence that hominids like *Homo habilis* were shaping stones. Hominids (*Homo erectus*) seem to have used fire for at least 700,000 years, perhaps for 1.6 million years or more (McCrone, 2000), although it is difficult to determine when the ability to make fire was acquired. Use of fire gave much improved protection from predators, the possibility of hardening wood spears, cooking food, and obtaining warmth during chilling nights and cold weather, and it may also have been used to smoke and preserve food for times of scarcity. It also very probably played a significant role in the development of human society as people congregated around hearths. Stone tools became highly developed and widespread: the Acheulian-type hand axe was one of the most widely distributed human tools between 1.5 million and 150,000 BP. Physiologically 'modern' humans appeared between 200,000 and 100,000 years BP. At what point language developed is difficult to determine – probably by 40,000 BP, possibly earlier. Another valuable innovation was the bone needle, which enabled humans to 'tailor' skin garments, improving their chances of survival in hostile environments.

Spears dating back to 420,000 BP or earlier have been found at Clacton and Boxgrove (eastern and southern UK), Bremen (Germany) and elsewhere, suggesting that hominids and humans were capable long before the last glacial not just of scavenging but also of hunting large mammals (Pitts and Roberts, 1998). In Europe during the past 35,000 or so years people have left well-preserved cave paintings showing game hunting, including elk, deer, mammoth and rhino being attacked with spears (rock paintings in Australia and Africa have been dated back to about 40,000 BP, and in southern Africa engraved ochre ornaments have been dated to 70,000 BP). In Eurasia semi-permanent settlements were constructed with mammoth bones and tusks during the last glacial in timber-scarce tundra areas. In Africa, Eurasia, North America and Latin America, kill and butchery sites have been excavated to reveal large game animal bones, sometimes with embedded flint spearheads and with cut marks.

There has been much disagreement about the so-called 'Pleistocene Overkill' theory – the suggestion that humans overhunted some species to extinction. Supporters point out that a wave of extinctions seems to match the likely pattern and rate of spread of humans: happening first in Africa and Southeast Asia, then Australia (in the latter from about 50,000 BP), Eurasia and, rather later, North America (starting around 11,000 BP), and finally South America. Roughly 85 per cent of the large mammals of the Americas became extinct during the last 11,000 years. The problem is that while some large mammals and their predators became extinct, others did not; it is also more likely that humans usually hunted easier game than mammoth or rhino. In southwest USA losses around 10,000 BP include mammoth and horse species, but not bison or moose. Analysis of ancient pack rat (*Neotoma* spp.) middens from that time, which are accumulations of plant fragments dragged into rat burrows where they are preserved by being soaked with uric acid, indicate marked vegetational change. The

Figure 2.3 European mammoth – hunted to extinction by humans, or victim of postglacial environmental changes?
Source: Oil painting (Plate 51 in *Prehistoric Animals* by J. Augusta and Z. Burian. Publication date unknown. Published by Spring Books, London, but printed in Czechoslovakia and translated into English by G. Hart). The plate is a reproduction of an original by Professor J. Augusta, exhibited in Prague Museum.

possibility is that large mammals like mammoth could not cope with altered grazing or the hindrance of shrub and tree cover as conditions warmed.

The pattern of large mammal extinctions might also reflect advances in killing technology (Martin and Wright, 1967; Martin, 1984). North America lost around 75 per cent of its large mammal species (over 45 kg weight) between roughly 13,000 BP and the present. Ager (1993: 192) noted the way a number of large mammals in western North America became extinct in a few hundred years (around 8000–7700 BP) when there were fewer marked climate changes than earlier, and was inclined to blame the development of Acheulian-type stone axes. Successive extinction of species may equate with development of tools to make fire, better wooden spears, spear throwers, flint-tipped spears, the adoption of the bow, and devices like the bolas. One technological advance that led to the near extinction of the American bison and many big game species in Africa, India and elsewhere, is established: the spread of rifles.

Another explanation has been offered for large mammal extinctions:

disease. There may have been natural epidemics, or wandering humans could unwittingly have spread viral diseases (Miller, 2001). Researchers have found signs of tuberculosis (TB) in a high percentage of North American mastodon bones. Climate change may have caused the extinctions, although those claiming anthropogenic causation seem to be gaining ground. Certainly, in the last several centuries, humans have been exterminating species, and that has been documented; the losses include Steller's sea cow (*Hydrodamalis gigas*) in Arctic and Alaskan waters in the eighteenth century, the Moa (*Diornis* spp.) in New Zealand, the passenger pigeon (*Ectopistes migratorius*) in the USA, several antelope species in Africa and India, wolves, beavers and bears in the UK, and in all probability several large marsupials and reptiles in Australia. Whether or not humans caused the extinctions they have had to adapt to changes in the availability of game species for most of the time they have been evolving.

The way indigenous people deal with the environment today, and have done in the past, has generated disagreement. In colonial days they were often seen as 'destructive', 'primitive' and at the mercy of nature. While some saw them as quaint, and others (like Rousseau) as 'Noble Savages', seventeenth- and eighteenth-century colonial governments often pretended there were no indigenous people, so land was *terra nullis* and there was no need to negotiate land or mineral rights. During the last century or so anthropologists and travellers have highlighted the skilled livelihood strategies and often sophisticated cultures developed by indigenous peoples. For the 1960s it was politically correct in many quarters to assume all past and present Aboriginal peoples were 'in tune with nature' and practised non-destructive sustainable livelihood strategies. Archaeology and anthropology often contradict these assumptions. Nevertheless, many hunter-gatherer societies have proved to be far from 'primitive' and have valuable lessons to offer modern humans. Indigenous people in a number of countries, once marginalized, now play an important role in land and natural resources development and are politically active (for a review of these issues see Krech, 2001).

As mentioned earlier, there have been a number of glacials during the past 2 million years or so, and, if humans had not caused global warming, the next would probably have started within 50,000 years. The last Quaternary ice advance (the Wisconsin) began roughly 30,000 BP and ended around 15,000. At its 'peak chill', *c.* 22,000 BP, ice covered about one-third of the world's land surface. Deglaciation took place between 11,000 and 9000 BP; around 11,600 BP there seems to have been an especially sudden and marked warming. From 12,000 to 14,000 BP, there was at times rapid melting; that of the North American ice sheets accumulated in lakes, which suddenly drained (this is believed to have suddenly reduced the salinity of the North Atlantic). At the same time there were increased numbers of icebergs; as these 'bursts' of icebergs melted they left telltale ice-rafted sediment on the ocean floor: **Heinrich events.** The salinity reduction either corresponded with or helped cause abrupt climatic deteriorations (cold stadials); one possibility is that these salinity reductions suddenly, and more than once, shut off the Gulf Stream and plunged Europe back into chill conditions.

Even during the postglacial the climate has fluctuated considerably, some of the better established changes (dates vary somewhat) are: the Older Dryas 'cold episode', starting roughly 13,100; a 'warm episode' (the Bolling); and the Younger Dryas 'cold episode', roughly 12,800 to 11,400 BP (Goudie, 1977: 35). The Younger Dryas seems to have almost been a return to glacial conditions, with a global fall in temperature of several °C, and it started and ended quite suddenly. The past 10,000 years have been very stable compared with most of the preceding million years or so.

Climate change during the last few million years has taken place at high and low latitudes. In parts of the latter there were periods drier than today and those that were wetter (pluvials). During the pluvials, many tropical lakes expanded and rivers flowed in areas that are now waterless deserts. Researchers have tried to establish whether climatic change was globally synchronous, and whether wet and dry phases at lower latitude equated with cold and warm conditions at higher latitude (Maarleveld and Van der Hammen, 1959).

Ocean temperature between 20°N and 20°S of the Equator probably remained fairly stable during the Quaternary. However, the idea that terrestrial conditions in the tropics changed little has been challenged by evidence from many disciplines. Geomorphological, palynological, archaeological and biogeographical (patchy modern plant and animal species distributions) evidence suggests shifting wet and dry conditions in the Amazon and Congo basins, and there were probably similar changes elsewhere in the tropics (some of these areas may have been attractive to humans during the glacials). A growing body of evidence points to cooling of tropical land areas rather than just rainfall changes. Certainly at higher altitudes in the tropics vegetation zones have moved up- and down-slope, and on high mountains glaciers advanced and retreated. Presumably the same changes were felt in lowlands. Palynology and chemical analysis of lake deposits from Amazonia indicate temperature falls during glacials of perhaps 5°C (Jones, 2001a). In Equatorial Africa and Southeast Asia during the peak of the last glacial, tropical lowlands may have had less forest cover than at the start of the twentieth century, the climate being cooler and drier than today's by between 5°C and 10°C (Flenley, 1979; 1998). The likelihood is that, briefly, parts of the Equatorial tropics experienced drier conditions, something that has happened to a lesser degree more recently during ENSO events (see Chapter 5).

There is evidence that the Kalahari and Sahara Deserts varied in extent during the last few million years. Pollen analyses from the Angola Basin (East Africa) provide a record back more than 300,000 years, suggesting that high-latitude glacial stages correlate with desert spread in that part of

Africa (Mannion, 1999: 46). Goudie (in Slack, 1999: 10–32) reviewed conditions in the tropics during the Quaternary and the implications this had for humans. Tropical glaciers expanded during glacials – for example, in Latin America snowlines in the Andes fell up to 1000 metres. Marine sediment cores suggest increased dust deposition at lower latitudes at those times. This implies drier and cooler conditions in the tropics during glacials; however, glacials and pluvials were probably not really synchronous. What is indicated is that weather systems have shifted; events like the Younger Dryas seem to correlate with ocean-atmospheric circulation 'flips'.

Ice core evidence shows that, in places like the centre of Greenland and Antarctica, temperatures fell 20°C or more below present levels at the peak of the last ice age. There are now ice core records with reliable annual information on climate and other palaeoecological data from a several parts of the world, which go back for over 100,000 years; there are also ocean and lake sediment cores and other sources covering the last 200,000 years or more. Recent research strongly suggests that for at least the last few hundred thousand years some of the climatic changes have been marked and quite sudden. Adams *et al.* (1999) discussed the many sudden marked climate changes during the last interglacial and found these worrying, because one can view that period as a close counterpart of the present. Some changes have been so fast that if similar occurred in the future, food production, national economies and transport would find it extremely difficult to adapt (Alley, 2000). Studies of fossil beetle remains in eastern UK by Professor G. Russel-Coope show that the postglacial was punctuated by rapid returns to cold conditions, possibly when the warming of the oceans or melting of North American ice sheets 'turned off' or shifted the Gulf Stream.

The most striking environmental change of the last 15,000 years has been the disappearance of continental ice sheets from Eurasia and North America (see Fig. 2.4). There has also been the shrinkage of ice cover elsewhere, and a rise in sea-level of as much as 390 feet (*c.* 130 metres) through the melting of the ice and expansion of oceans as they warmed. Between 14,000 and 10,000 BP – as the glaciers retreated from Europe, Eurasia, North America and South America – tundra gave way to steppe and then to a succession of forests. Between 10,000 and 6000 BP, during the 'Holocene Maximum', the Upper-Palaeolithic, global temperatures were about 2°C above those of the present day and rainfall patterns were different. Large parts of what is today dry Saharan Desert and some Asian drylands were relatively well-watered savannahs. In the Sahara, for example, there were crocodile and hippo in regions now far from any suitable water body, and areas like the Tibesti Massif were inhabited, but today these are far too dry.

Humans were probably forced to withdraw from what today are African and Middle Eastern drylands and arid parts of Eurasia and Southeast Asia around 4000 BP as rainfall patterns changed. At about the same time rising sea levels also probably caused relocations and increasingly hindered migration on foot. It has been suggested that drier conditions prompted

a)

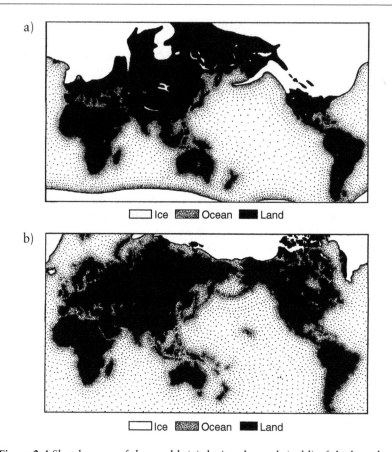

b)

Figure 2.4 Sketch maps of the world: (a) during the peak (cold) of the last glacial *c.* 18,000 years BP, and (b) the present day
Note the obvious difference in ice cover (the white shading) and the greater exposure of land during the glacial due to lower sea-levels in the Mediterranean, around India, in Southeast Asia, around Japan and Korea, and in the Caribbean.
Source: Reproduced and modified with permission from Fagan, B. (1999) *Floods, Famines and Emperors.* Pimlico Press, London.

movement to river valleys in Mesopotamia and those of the Nile and Indus, and shrinking coastal lowlands also helped stimulate agricultural and technological development and less scattered settlement. Large towns and cereal growing appeared in northern Iraq and Palestine (the Natufian Culture) by 11,000–9000 BP. By 7000 BP domesticated sheep, pigs, cattle and goats had reached Greece, and by 7500 BP copper was being smelted in southern Europe.

In Europe and elsewhere there is evidence for a 'climatic optimum', with an average world temperature 2°C to 3°C above present, between roughly

7000 BP and 5000 BP. Conditions seem to have started to become cooler between 5400 and 5000 BP, and in Europe there are signs of wetter conditions from 3250 to 3200 BP, and also between 2900 and 2600 BP, and 2400 and 2100 BP.

2.3 Changing sea-levels

Throughout the Earth's history sea levels have varied as ice caps have expanded or shrunk, as seas have warmed or cooled (and hence expanded or contracted), and as ocean basins have altered in size through tectonic changes and continental drift. Over the timespan affecting human development, the first two causes have been responsible for sea-level changes. During glacial maxima the oceans fell more than 130 metres below present levels, and during the warmest parts of interglacials (before 110,000 BP) probably rose over 70 metres above present sea-level. Sea-level rises between 13,000 and 4000 BP have caused considerable impacts. Today, there are extensive, relatively shallow seas around many continents, which would have been exposed as dry land for long periods before the end of the last glacial. The North Sea still had extensive land areas until *c.* 3000 BP. Some of these areas are so large that their exposure and drowning may have played a part in climate change. With humans driven from the higher latitudes by glacial conditions, now submerged tropical and subtropical areas are also likely to have hosted a significant part of human development (Oppenheimer, 1999).

2.4 World flood myths

It is striking how many cultures have a tradition of past catastrophic, often worldwide, flood (including the Cherokee Indians of the USA, Indian peoples in the northwest of North America, Polynesians, the Akkadians, ancient Greeks, peoples of Southeast Asia, India, Sri Lanka, Japan and many others). These myths are often strikingly similar, which is hardly surprising given the marked sea-level changes of the last 15,000 years, plus occasional tsunamis. The latter may have been more frequent in the period 15,000 to 7000 if earthquakes were generated as a consequence of the loss of the huge weight of ice on higher-latitude landmasses. Also, tsunamis can be caused by mass movement of submarine sediment down continental slopes, and such landslides may have been more common 10,000 to 6000 years ago when there was plenty of unconsolidated sediment that had been deposited during the glacial periods, and that was being disturbed by warming seas, tectonic uplift and associated earthquakes. There may also have been more mass movement activity when sea-levels fell during glacial periods.

Roughly 2400 BP the Greek philosopher Plato recorded the story of Atlantis (in *Timaeus* and *Critias*, which he apparently derived from ancient Egyptian sources), drowned around 11,600 BP. Celtic legends tell of the lost land of Lyoness, suddenly submerged somewhere off the southwest of England; the Judeo-Christian tradition has the story of Noah's flood (Genesis: Chapter 8), which shares elements of the, probably earlier, Babylonian Gilgamesh Epic (recorded *c.* 660–630 BC, and thought to tell of events around roughly 3400 BC).

There was a marked worldwide Flandrian transgression (a relatively rapid sea-level rise) about 11,600 BP; Oppenheimer (1999) and Hancock (2002) argue that there are indications that some cultures had already developed cities and sophisticated civilizations by that time, when they were inundated. Off southern Norway about 7000 BP, a mass movement of sediment down the continental slope (the Storegga landslide) sent tsunami waves *c.* 700 km to deposit sand across peatbogs in northeast Scotland up to 30 above present sea-levels (with lower sea-levels at the time this may have been quite a large tsunami), and it must have struck elsewhere around the North Sea. Some geomorphologists suggest that the Storegga tsunami helped separate the Orkneys from the British mainland, and the British Isles from Europe.

The problem is not so much finding past flood events, but of reliably matching them to the legends; also, myths like that of Atlantis and Noah's flood have attracted cranks, incautious environmental historians and archaeologists, and assorted 'pyramidiots', deterring more rational and thorough researchers. Underwater archaeology is in its infancy, and marine submergence or tsunamis would destroy much of the more obvious evidence.

There is no shortage of explanations for the Atlantis myth; recently some have become more rational (James, 1995). Oppenheimer (1999) tried to assess objectively whether there may have been actual events behind the myths. He focused on the flooding of the huge shallow areas off Southeast Asia (the Sunda and Sahul shelves). During the last glacial these were a huge lowland area of over 2 million km², drowned somewhere between 17,000 and 7000 BP (Hantoro *et al.*, 1995). These lands may have supported flourishing cultures when higher latitudes were inhospitable; if so, the flooding could have caused a population dispersal, perhaps to China or South Asia, possibly Egypt and the Middle East, and even initiated Polynesian movements. In early 2002 radiocarbon dating of wood samples recovered with numerous artefacts and human bones from the Gulf of Cambay off Gujarat (seaward of the Indus River), together with sidescan sonar surveys suggest extensive settlements and a sophisticated culture, possibly before 9500 BP. A journalist, Graham Hancock (2002), has tried to stimulate research to establish whether human civilization is much older than currently acknowledged and has had to repeatedly move and adapt. His hypothesis, that civilization developed in areas below present sea-level at least during the late glacial when higher latitudes were too cold, has been greeted with much opposition from established disciplines. The ideas deserve consideration; there are huge now-drowned areas that would have been attractive during the last glacial until they were inundated by sea-level rises around 14,200, 11,600 and 7600. There is also a chance that large tsunamis were relatively frequent during deglaciation.

What are believed to be flood deposits have been found at Ur and other Sumerian cities (Sumer Valley, Mesopotamia), roughly dated to around 3000 BC; these might relate to the Gilgamesh and Noah legends. Another possibility, which Ryan and Pitman (2000) have examined, relates to Bronze Age settlements lying *c.* 100 metres deep in the Black Sea, about 20 kilometres off the present Turkish coast. These settlements, and the surrounding lowlands and lakes, probably flooded about 7500 BP (*c.* 5800 BC), when postglacial sea-level rises breached a land barrier across the Bosphorus. This flood might have driven some people south (carrying the Noah legend) and other refugees into the Danube Valley.

Many dismiss Plato's Atlantis legend as a fabrication or an allegory, others trace it to recollection of the loss of a Greek city on the Gulf of Corinth, or the explosive eruption of Thera and the destruction of the Bronze Age Minoan civilization on Crete and other islands of the eastern Mediterranean (Friedrich, 2000). A group of geoarchaeologists suggest the loss of a Mycenaean city – Tiryns – might have prompted the legend (Zangger, 2001). Others follow up Plato's mention of the Atlantic, and seek sites and explanations anywhere from the Azores to the Caribbean, including the Canary Islands (recent speculations centring on an Andean lake seem less likely). French researchers who accepted that the location given by Plato as just beyond the Pillars of Hercules (close to the Straits of Gibraltar) was reliable, suggested that Atlantis lay in the Atlantic approaches to the Mediterranean, between Spain and Morocco, where coastal lowlands and an archipelago (now the submerged Spartel Bank) flooded about 9000 BP (*New Scientist* 22 September 2001: 17).

Humans probably have vague and garbled memories of catastrophic floods, probably related to postglacial sea-level rises or tsunamis. Planners should take heed of past events and review the siting of irreplaceable or especially sensitive installations, such as libraries, gene banks and nuclear waste repositories.

Key reading

Adams, J., Maslin, M. and Thomas, E. (1999) Sudden climate transitions during the Quaternary, *Progress in Physical Geography* 23(1), 1–36.

Alley, R.B. (2000) *The Two Mile Time Machine: ice cores, abrupt climate change and our future*. Princeton University Press, Princeton (NJ).

Burroughs, W.J. (1997) *Does the Weather Really Matter? The Social Implications of Climate Change.* Cambridge University Press, Cambridge.

Fagan, B.M. (1995) *People of the Earth: an introduction to world prehistory* (eighth edition). HarperCollins, New York (NY).

Turekian, K.K. (1996) *Global Environmental Change: past, present, and future.* Prentice-Hall, Upper Saddle River (NJ).

|3|

Environmental change during historical times

From the jungles of Guatemala to the shores of western Greenland fjords there are the ruins of societies which seem to have been weakened by the effects of prolonged adverse weather.

(Burroughs, 1997: 1)

Chapter summary

This chapter explores environmental change from roughly 4000 BP to the present. Arnold (1996: 1–2) noted that

> Many historians belong to a tradition, still in many ways the dominant one, in which nature, as either ideology or material reality, does not figure, except perhaps to set the scene for the real drama to follow: the drama of human lives, human action, human-centred events. ... But of late ... historians have begun to take a far more positive attitude toward nature's place in the writing of history ...

Two major differences distinguish this from the last chapter: the availability of written reports or records, and the increasing influence of human activity on the environment.

Many of the word's cultures have calendars and celebrate festivals strongly influenced by the seasons. The phases of the moon were (and often still are) widely felt to affect crops, fishing, animals, love, fortunes and the mental state of people. Some of the first marks scratched by humans on bones or cave walls probably record seasons and natural changes like the phases of the Moon. In spite of this deep influence of nature, conventional approaches to history are anthropogenic in outlook and neglect to adequately consider interactions with the environment. However, there are growing numbers of environmental historians who seek to identify and understand past interactions between environment and humans. There has been a tendency to focus on climate, although this is only one of many environmental parameters affecting human fortunes (Ponting, 1991; Bradley and Jones, 1992). To make reliable assessments about past environment–human interactions is not easy (people deploy social and technical strategies to adapt better and this can often be difficult to study); often correlations have been speculative and unproven.

The way human behaviour responds to environmental changes is not linear and constant; there are periods of stasis, sudden shifts, setbacks and recurrent cycles (Worster, 1988: 179). This was something noted by N.D. Kondriatieff (1892–1938) in the 1790s: human affairs showed a 'long wave' pattern reflecting the time it took to relate to the environment and to make use of innovations (see Fig. 3.1). For example, a group of people may resist bad weather for a few years, and then suffer if difficulties continue. A climate change can also have a selective impact: cool conditions might kill more older people, and warmer conditions cause more deaths of young children (through gastrointestinal infections).

Evidence of ever-improving accuracy, detail and variety, comes from many fields (Crowley and North, 1991). There is also written (historical) evidence: in Europe chronicles and parish registers sometimes give insights back to AD 700, and Chinese dynastic records and written evidence from Egypt can provide evidence of environmental conditions as far back as 3000 BC (Bradley and Jones, 1992: 6). Often there is a suspicion that environmental factors have determined human affairs, but it cannot be proven because the evidence is too imprecise, and because environment–human interactions are rarely direct, simple and linear. Figure 3.2 illustrates (in very much simplified form) the indirect, often cumulative, impacts and complex feedbacks that can be encountered when dealing with just one environmental parameter; things get far more convoluted if an attempt is made to factor in anything more than a fraction of real-life complexity.

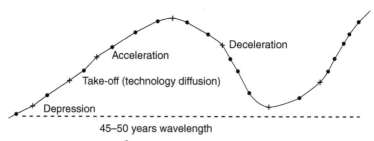

Figure 3.1 Kondriatieff[*] long wave
Note: *'Kondriatieff' can also be spelled 'Kondratyev'
Source: Diagram drawn by author

Historical evidence can take a variety of forms, including the following.

- Recordings of actual observations of weather, volcanic eruptions, storms and so on. These are unlikely to go much further back than the seventeenth century AD.
- Mention of conditions or events related to weather, eruptions, storms, etc. Descriptions of 'frost fairs' indicate that London's River Thames had thick ice in AD 1309, 1564, 1608, 1684, 1693, 1715, 1739, 1789, 1811 and 1814 (Currie, 1996). These records must be interpreted with care; before 1831 the Thames almost certainly froze more easily because the Old London Bridge slowed the flow, also there was little of today's waste heat from factories and power stations. Nevertheless, there is no mention of frost fairs between AD 1434 and 1540. There are records of the Baltic Sea freezing in AD 1296, 1306, 1323, 1408, 1423, 1428 and 1460. Also, in Europe, Church documents sometimes record rogation ceremonies, which were called to pray for rain in time of drought or an end to other adverse weather conditions.
- Phenomenological records, such as the date ice breaks up or swallows arrive; of invasions of locusts, dates of sowing or harvest; stories of winters severe enough to drive wolves into areas normally free of them, mention of wells and springs drying up.

Sources of historical data are thus diverse, but can be broadly split into the following groups.

1. Quite factual documents that sometimes have accurate, even quantitative, information (e.g. observatory records, state archives, shipping records, the logbooks and journals of sea captains or trading posts – some like those of the East India Company or Hudson's Bay Company go back to the 1750s or earlier).
2. Archives that, although not rigorous quantitative observations, can be strongly correlated with environmental conditions (e.g. agricultural records of quantities of grain or wine harvested and areas sown).
3. Wholly qualitative and less precise mention of conditions (e.g. correspondence, diarists). Unfortunately, a qualitative mention of something like 'drought' may convey little on its actual severity relative to other similar events; a strikingly cold winter may be recorded, but have less effect on human affairs than a following dull wet summer or a brief unseasonable frost, which escape mention. Bradley and Jones (1992: 147) relate how locust invasions of parts of Europe via the Danube might be incorrectly interpreted as an indication of warm summers, whereas they are more likely to result from more intensive land use in Hungary and possibly unusual wind conditions.

Ongoing records are sometimes kept by different people who vary in reliability or the stress they place on events. Some records document conditions quite well, but omit details of exact date or location – the latter is a

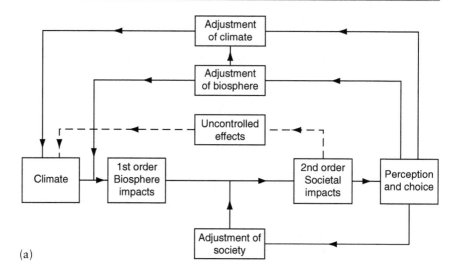

(a)

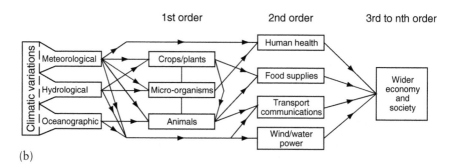

(b)

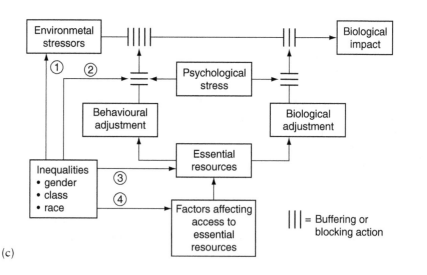

(c)

particular problem with ships' logs until it became possible to determine longitude accurately in the late eighteenth century. Interpretation is also crucial: studying western European records, Burroughs (1997: 32) found correlations between high wheat prices and increased infant mortality a year or more later; his assessment was that damage had been done during pregnancy and that the impact of climate was delayed, and also that social and economic factors alone or in combination with environmental causation could be responsible. Often, various researchers might make different assessments of the same data.

Cautious interpretation is needed; farms and settlements may be abandoned as a result of many things as well as climate change, including warfare, diseases like smallpox or other plagues, exploitation by landlords, national tax levies, out-migration and economic recession (Beresford and Hurst, 1971; Parry, 1978). However, there does appear to be a fairly common recurrent pattern: bad weather (in higher latitudes, cold or wet; in lower latitudes, drought or floods) causing famine and an increase in disease vector organisms; epidemic, followed by labour shortage and leading to degeneration of agriculture; and then further debilitation and disease – a downward spiral growing ever tighter. Such a sequence can still be seen in poor countries, so the present can offer some clues to the past.

3.1 Blaming human history on environmental change

Investigations should rely on as many lines of evidence as possible and all interpretations must be very cautious. Some have tried to read too much from evidence of limited reliability, while others have made meticulous studies to assess past conditions. An example of the latter is the historical climatology of Sir Hubert Lamb (1977; 1982; 1988), Jan de Vries (1980) and the work of the thorough environmental historian Le Roy Ladurie (1972).

Fig 3.2 Climate–human interactions
(a) Conceptual model of the impact of climate on humans and society, and possible feedbacks via adaptive strategies. Uncontrolled efforts are inadvertent and unplanned.
Source: Reproduced with permission from Wigley *et al.* (1981) *Climate and History*. Cambridge University Press.

(b) Effect of climate on society. As the impacts become more indirect (2nd to nth order) the chain of causation is more complex and difficult to trace.
Source: Reproduced with permission from Wigley *et al.* (1981) *Climate and History*. Cambridge University Press.

(c) A political economy model of environmental constraints, showing how inequalities intensify exposure and reduce adaptive potential.
Source: Reproduced with permission from Ulijaszek, S.J. and Huss-Ashmore, R.A. (eds) (1997), *Human Adaptivity*. Oxford University Press, Oxford, New York.

There has been considerable, often heated, debate about the extent to which past human migration is related to environmental change. A number of environmental historians have suggested this is the case, including Huntington, Claiborne, Toynbee (1934; 1976) and Markham (1942). Around roughly 4000 BP a number of civilizations seem to have 'collapsed' in Greece and the Aegean, Egypt, the Persian Gulf, Mesopotamia and north-west India; this tempted environmental historians to blame climatic change (toward drier conditions). However, *proving* this is difficult. Pearson (1978: 211–29) urged caution in attributing human actions to environmental causation, yet was prepared to concede (1978: 213) that 'overgrazing can have disastrous results under any circumstances but it is clearly more likely to do so during a period of climatic deterioration which reduces even further the population of ungulates that a given area can support'.

Looking at past environments it is important to distinguish between the occasional extreme event, some random, but others showing periodicity, and actual shifts in climate, i.e. altered average conditions. There has been a tendency for 'cold' periods like the 'Little Ice Age' (LIA) (see later on in this chapter) to be seen as 'abnormal' and recent conditions as 'normal', a dispassionate examination of the last few millennia should cause one to question such attitudes. Similar views are often encountered in regions with variable rainfall: wetter than 'average' (a rather meaningless term in such situations) years encourage exploitation and tend to be seen as normal; drier years are then viewed as abnormal. Also, there is a general feeling that change has usually been, and will in the future be, quite gradual; this is not a wise assumption. Modern humans should study the fortunes of people like the Norse (see Box 3.1) and other cultures that appear to have been affected by environmental change. Future environmental changes will cause less misery if people are watchful, flexible and adaptive. I worry that the people of today's developed nations are much less watchful, flexible and adaptive in times of stress than their ancestors. People of poorer countries may also have less opportunity to adapt than in the past, because of population growth, loss of access to common resources and other negative developments. Modern humans have less leeway, in spite of better technology (there are now over 6.5 billion to feed; in AD 1500 there were only around 400 million people affected by the Little Ice Age). Fagan (1999: 91) made the same point in relation to drought: 'Dense urban populations in circumscribed environments require highly productive agriculture just to survive, let alone make it through drought cycles or natural disasters.'

One might argue that humans have become more vulnerable to environmental problems since starting to shift to agriculture. This has made people much less mobile and robbed them of most of their skills in surviving off the land; worse, populations have increased too much. The shift to agriculture and modern lifestyles has generally been seen as progress, yet researchers like Brody (2000), who has studied the Inuit, argue that some hunter-gatherers live in harmony with their environment, in demographically stable

Box 3.1   The Norse settlement of Iceland, Greenland and 'Vinland' – the difficulty proving environmental causation

The settlement of Iceland and Greenland, like that of Easter Island in the mid-Pacific (for a short review of the latter see Goudie, 2002: 310–14), and the short-lived English colony at Roanoak in North America (1587–89?), has attracted environmental historians, hoping to learn lessons relevant to modern development. Greenland is an interesting case study, the widely held view being that because the settlements were close to struggling even under optimum climatic conditions ('subsistence on a knife edge'), any adverse environmental changes would soon be apparent.

During the Medieval Warm Period, Iceland – along with the Faroe Islands, Shetlands, Orkneys, northern Scotland and the Isle of Man (UK) – were settled by the Norse from Scandinavia (there may have been even earlier settlement of Iceland by Irish Anchorites). Between about AD 870 and the onset of the LIA farmsteads and villages were established and churches built where the Icelandic environment was favourable (Bradley and Jones, 1992: 92).

In the tenth century AD, having committed murder or become involved in a power struggle with other chiefdoms, Eirik ('the Red') was outlawed and banished from Iceland. Icelandic sagas (epic stories) tell how he went and founded colonies on the southeast and west coasts of Greenland between 982 and 986 (earlier Norse settlement may have been attempted around 930). Palaeoecological studies by Fredskild (1973) show that Greenland had warm but variable conditions between AD 900 and AD 1130; there was cooling from the mid-twelfth century until about AD 1380; the period AD 1550 to AD 1850 was cold (the LIA); and AD 1880 to the 1950s was a warming phase, with post-1950 some further rise to present temperatures.

For 400 years the Greenlanders established farms, kept cattle and grew crops in areas that are now icebound or frozen tundra. Archaeology suggests that before the twelfth century AD graves were dug deeply and plant roots intruded well down into the soil; but later, there were shallow burials with roots restricted to the topsoil – signs of permanently frozen subsoil. At the peak of settlement, around AD 1126, Greenland had at least 16 parish churches, and a new cathedral at Gardar was sent a bishop by the Vatican in AD 1127. At that point the population is estimated to have been around 3000 people.

With North America as close to the west as Iceland was eastward, it was inevitable that the Norse would discover the New World either on deliberate explorations (Greenland had no building timber) or when ships en route to and from Greenland were blown off course. Sagas tell how a merchant, Bjarni Herjolfsson, sighted what might have been Labrador or Newfoundland in 985 or 986, and there are other mentions of landfalls west of Greenland, like that led by Eirik's son Leif around 1001, together with vivid descriptions of timber cutting and encounters with Indian peoples. Between the tenth and fourteenth centuries AD the Norse had a chance to settle America. Had there not been a decline of the Greenland settlements, larger groups might have moved west and been better able to compete with the Indian peoples. One can speculate that, were it not for the LIA, the Norse may have settled the New World five centuries before Columbus. Although McGovern (in Crumley, 1994: 127–54) is more sceptical, arguing that America was just too remote and, that while post-1492 settlers from southern Europe carried diseases that helped subdue the indigenous peoples, the Norse were relatively disease-free.

**Box 3.1   The Norse settlement of Iceland, Greenland and 'Vinland' –
the difficulty proving environmental causation – continued**

The Greenland settlements seem to have ceased flourishing after 1260, with the final decline under way by 1350. Norwegian records tell of shipping from Greenland to Iceland having to veer further southward by 1250 because of increasing sea ice. This would have discouraged contact at a time when Europe also had famines (1315–22). West Greenland was probably evacuated between 1341 and 1363, and the eastern settlements a little later, around 1500. Archaeologists favour the idea that most Greenlanders abandoned their settlements because of environmental factors; there is little evidence of conflict with the Inuit (Thule people) who moved southward between AD 1000 and 1600 (Wigley et al., 1981: 422).

Between AD 1270 and 1390, cereal growing in Iceland declined, ceasing after the second half of the fourteenth century, and it was not until recent times that production resumed. Records show that Iceland was surrounded by sea ice except for one port in AD 1695. The problems in Iceland and Greenland seem to coincide with the abandonment of vineyards in England and Wales, and the shift in weaving woollen cloth (which demands certain environmental conditions) toward the west in England. Hardship in Norway may have reduced trade with Iceland; the population of Norway seems to have been slow to recover after the ravages of the Black Death in the fourteenth century compared with Sweden and Denmark. Whether some climatic shift hit maritime environments like Norway more than other areas is not clear. Careful assessments of the effects of the Hanseatic League on trade suggests that Iceland's difficulties were environmental rather than economic (Worster, 1988: 42). In the early twentieth century, the Swedish oceanographer Otto Petterson argued that tidal variation periodically caused shifts in the North Atlantic climate, and that it was this that had finished the Greenlanders – a hypothesis that has attracted little further study (Worster, 1988: 42).

From 1410 Greenland had little or no contact with Europe or Iceland, and communication ceased entirely from about 1476. Various researchers have speculated that the remaining Greenlanders left or died out between 1475 and 1500. Conditions eased in Iceland between 1500 and 1550, and attempts were made to re-establish contact with its old colony; however, an expedition in 1540 found no survivors (Pringle, 1997). After 1550, sea ice spread again and contact ceased until 1822.

The decline might have been linked to land degradation as well as climatic deterioration; excavation of burials suggests that in better times settlers' heights averaged 5 feet 7 inches (168 cm), and that by AD 1400 this had fallen to 5 feet 3 inches (158 cm). This evidence is contested, but might be indicative of sustained malnutrition, reflecting overgrazing and excessive fuelwood collection, or the spread of poor-quality grasses, as much as climatic deterioration. Climatic historians have argued that westerlies increased in the mid-twelfth to thirteenth centuries, and records note that cool and wet summers prevented the growing of grain and hindered hay-making for vital winter fodder. People living 'close to the edge' of subsistence can easily succumb to a bad winter, a storm, drought or other natural disaster, and many of today's poor live such a life. There must have been some environmental deterioration because even now agriculture in Greenland is virtually impossible.

Excavations at Nipaatsoq (in western Greenland) suggest that glaciers expanded during the LIA and in summer their meltwater often flooded farmland and deposited

**Box 3.1 The Norse settlement of Iceland, Greenland and 'Vinland' –
the difficulty proving environmental causation – continued**

infertile sands. The settlers seem to have made some adaptations, weaving animal fur into their cloth, eating more seals and fish, and building single farm buildings rather than a cluster so that humans and livestock could share some warmth. There are indications that in that region at least, as conditions deteriorated people finally evacuated. Spreading sea ice would have hindered, and later possibly prevented, communication with the outside world and fishing probably suffered when stocks of Atlantic cod (*Gadus morhua*) and other important fish moved south. Cod are sensitive to water temperature – favouring that between 2°C and 13°C – and there was a marked movement of cod northward during the twentieth century in response to the sea warming.

The Greenlanders may have suffered from climate deterioration because they failed to adapt their dress and livelihood strategies sufficiently. Shipbuilding may have declined as sea ice reduced opportunities for collecting timber; and the Norse appear not to have made skin-covered boats like those used by the Inuit, and their winter seal-hunting skills, harpoon technology and superior-tailored skin clothing were not copied. These failings may have severely limited European Greenlander's food supplies, while the Inuit survived. Other factors might have included high taxation, the conservatism promoted by their traditions and clergy, lack of timber for shipbuilding and household use, and isolation (Pain, 1994).

Other factors that must be considered are disease, and trade. Western Europe was repeatedly struck by the Black Death and smallpox from the eleventh century AD – the former hit Iceland in 1402/04, and the latter in 1707/08. In 1261, Greenland lost its independence to Norway and trade was controlled. There was already at that time a decline in the demand for Greenland furs, as a consequence of trade opening up with Russia, and a fall in the popularity of another crucial export – walrus ivory – as better-quality elephant ivory reached Europe from Africa. There may also have been more opportunities to settle or plunder Europe than there had been when the Norse first settled Iceland and Greenland. Records suggest that the last official trading ship to sail to Greenland may have been in 1367 or 1369. McGovern (in Crumley, 1994: 148–9) suggested that the failure of the Greenland Norse to survive environmental challenges was cultural – they did not adapt their trading, missed the opportunity to adopt Inuit livelihood strategies and fell victim to poor management. As conditions deteriorated, the response may have been for smaller farmsteads to seek alms from larger Church-owned farms and richer landlords, rather than turning toward the sea and exploiting more marine resources like the Inuit.

Settlers on Iceland were subject to volcanic eruptions, floods, plagues and worsening climatic conditions after about 1550. They survived, although land abandonment continued well into the eighteenth century. The Icelanders made a shift from wheat growing to sheep farming during the LIA. Whether this was caused by environmental conditions or labour shortage following epidemics is uncertain. Iceland's population fell during the LIA to half what it had been during the preceding Medieval Warm Period.

Clearly, the environment presented challenges; why the Greenlanders failed to adapt is attributed to several possible causes. As archaeology and palaeoecology gathers more evidence the picture should become clearer.

Sources: Jones, 1964; Magnusson and Palsson, 1965; Bryson and Murray, 1977: 71; Fredskild, 1973; Pringle. 1997; Crumley, 1994: 127–54.

communities that are free from infectious disease, and in relative 'prosperity' with more time for leisure, skills development and cultural activities than many who practise agriculture. It should also be noted that hunter-gatherers may flourish in environments that are just too inhospitable for agriculturalists. It must be stressed, though, that in today's environment hunter-gatherers may not be like those who in pre-agricultural times were able to occupy less harsh lands.

The late sixteenth century was a time of unusual hardship, often apparently followed by unrest: Europe had severe famine in 1591, and between 1596 and 1598. Hunger returned to mainland UK in the 1690s, when there were many deaths. There were crop failures in France in 1709 and 1771, and the harsh winter of 1788–89 might have helped spark peasant unrest prior to the French Revolution in the summer of 1789. Archaeologists and historians can cite many examples that *appear* to provide evidence of environmental change affecting human fortunes (see Box 3.2). Hunger does not, however, automatically lead to rebellion. It is difficult to prove that fields became salinized because of drought; it might equally well reflect

---

Box 3.2 The English Industrial Revolution: possible human–environmental–technological linkages

A number of historians have tried to identify the 'triggers' for the English Industrial Revolution after AD 1750. The following factors have been noted by various historians, some of them environmental.

- *Reduced availability of human labour as a result of the Black Death.* Population in Europe seems to have recovered to around 80 million by AD 1500, and to have risen to 140 million by 1750, and to 220 million by 1850. So shortage of labour seems unlikely to be a cause.
- *Land shortage due to rising population in the sixteenth and early seventeenth centuries.* The response seems to have been technological innovation in already established industries to offer alternative employment.
- *A rising population led to wood shortages.* Consequently coal was substituted.
- *With few new sites for watermills, and because demand for coal and metals had led to deeper mines, technological innovations took place.* Steam pumps had begun to replace rag-balls-and-chain pumps and other primitive water-lifting devices that could not cope with deep mines by the 1690s.
- *Local areas could not supply their resource needs so canals, turnpike roads and railways developed.* Speculation in this infrastructure by investors played a significant part.
- *Wool and linen production demanded land that could be used for food production, so substitution by cotton imports began.* Land no longer needed for flax and sheep could be used for food production.

Note: There had been enclaves of industrial activity before the 1750s in regions like the Weald of Kent (southeast UK) even in Elizabethan times (by the 1560s). The Weald had iron smelting, cannon making and gunpowder manufacture on a large scale – but using local resources.

insufficiently careful agriculture or a gradual process of soil degradation. The abandonment of farms or a settlement may share a date with a drought or a volcanic eruption but correlation does not *confirm* causation – there might have been simultaneous political or social change. Plagues have struck people at various times throughout history and may be triggered by climatic or other ecological changes; interestingly, rich and poor, urban and rural folk have all been hit by some outbreaks, so diet and lifestyle can be ruled out in those cases, making environmental causes more likely (Gottfried 1983; Arnold, 1996: 67–70).

The impact of environment upon humans depends on how well society and the economy adapt. Because adaptation can follow a diversity of paths, interpreting past events can be difficult; for example, where people increase their insurance claims over a period of time, events like storms may seem to have become more severe, but in reality this may reflect changing settlement patterns, trends in taking out insurance, and the rising cost of home furnishings and building materials. Even if some past environmental changes can be proven to link with human fortunes, this may not offer lessons relevant to the world today and in the future, because human societies and economies have altered. Burroughs (1997: 1) made this point, noting that people in the twenty-first century might not respond in the same way as those of earlier times.

Agricultural technology can reduce vulnerability to environmental conditions – for example, making it easier to plough enough land without waiting for suitable weather (more primitive agriculture may fail to plough enough in time). Irrigation is widely used to improve security against drought. However, modern agriculture, with high-tech ploughs and tractors, can still be disrupted (in 1969 large parts of the eastern UK were left unplanted because of a wet spring). In 2000/01 wet autumn and winter conditions similarly prevented the planting of winter cereals over huge areas, and the following year livestock in the UK were decimated by foot and mouth disease.

The impact of severe weather can vary a great deal with altitude, soil type, aspect and so on; settlements just a short distance apart may be affected very differently by the same event. Damage can be caused by unseasonable or severe frosts, unusually long and cold winters, a wet and cool summer, a warm and wet winter or a warm winter that fails to kill crop pests and human or animal disease vectors, or an unusually hot and dry summer; these may have very selective impacts, perhaps striking only certain land uses or activities. It is also important to consider the full (indirect) implications of severe weather – for example, heavy rainfall may puddle soil and reduce infiltration for a long time afterwards, possibly causing enhanced erosion.

During the British Raj (the colonial period in India) there was considerable effort to understand, model and predict the South Asian monsoon rainfall in the hope that it would be possible to prepare for 'unexpected' crop failures and famine. H.F. Blandford, a senior officer in the Indian Meteorological Service, tried to relate winter snow cover in the Himalayas to the following monsoon strength, with little success (although these studies helped initiate research that later exposed the ENSO phenomenon; see later on in this chapter). Some 150 years later researchers are still looking at links between climate fluctuations and famine in India, Ethiopia, the Sahel and elsewhere. Are today's administrators any better prepared than their predecessors over the last 3000 years in terms of contingency planning and stockpiling of food reserves to withstand climatic variation and severe weather events, let alone the possibility of longer-term changes (e.g. a long-term rainfall or temperature shift)? Apart from improved transportation, probably not. Yet, as Bryson and Murray (1977: xiii) warned, 'Climate variation, like death and taxes, is certain. We know of no century with constant climate.'

In the Northern Hemisphere there were warm conditions between about 5300 and 4600 BP, the Holocene Optimum, and from *c.* 3200 to 2500 BP conditions in Europe cooled. From about 4500 to 4000 BP the cities of Harrapa and Mohenjo-Daro (Indus Valley) declined, perhaps due to a reduction in rainfall. Around the same time (4500 to 4000 BP), civilizations in the eastern Mediterranean, including the Hittites, Mycenae and Ugarit, seem to have collapsed as a consequence of their territory being invaded by warlike settlers (Burroughs, 1997: 17). Also, the decline of the Classical Greek 'Golden Age' took place at about this time. Were these misfortunes caused by the reduction of summer rainfall and cooler winter conditions east and north of the Mediterranean triggering movements of invaders?

The Mayans established an impressive culture in sub-tropical Central America (Meso-America); excavations of their cities and evidence from lake sediment cores in Yucatan (Mexico) indicate that the Mayan culture – which flourished for around 3000 years, reaching its peak for about eight centuries (roughly AD 250–1200), growing to about 11 million and establishing over 40 known cites – collapsed relatively quickly, and by AD 900 was a shadow of its former glory. Given the length of time the Mayans sustained one of the densest populations of the ancient world, it is interesting to try and establish the cause of the collapse around AD 900. With no apparent signs of social unrest or external threats the temptation has been to seek environmental causes, something originally suggested by Huntington in 1915. The volcano El Chichón, which erupted in AD 900 and 1250, has been blamed for the demise of the Mayan civilization by some researchers. However, more recent studies suggest a decline attributable to severe droughts, possibly related to El Niño events, which the people were unable to adapt to, possibly because rising population had degraded the land (Burroughs, 1997: 1; Faust, 2001) (see Chapter 5 for a discussion of El Niño, La Niña and ENSO). Sceptics have argued that ENSO-related drought is unlikely in humid tropical Yucatan; yet the Mayans constructed sophisticated rainwater harvesting structures, which shows they had problems. Records in Mexico report severe droughts several times since the Conquest, one in Yucatan in AD 1795; and in 1997/98 there was an ENSO-related reduction of rainfall in usually humid Panama, which troubled the Canal Authority with too little water to operate the locks at normal capacity. Patterns of rainfall can shift, even in the humid equatorial tropics. In various parts of the Amazon basin soil cores show charcoal layers which indicate that, at times, the forests have dried enough to burn, possibly during serious ENSO-related droughts.

The 'collapse' of the Chaco Anasazi Culture on the Colorado Plateau (southwestern USA) and that of the Moche Culture in western South America have also been blamed on droughts. The Anasazi established many settlements between AD 159 and 490, and prospered until around AD 1299. Their territory was affected by drought between 1000 and 1180 AD, and again between 1275 and 1299, after which they seem to have dispersed. In North America the middens of long-abandoned Indian villages show repeated shifts from deer to bison bones between AD 900 and AD 1400, interpreted as indicating variation from dry to wetter conditions, which caused game animals to alter their range (caution is needed, because some reason other than drought might be behind such changes). Drought may have hit the southwestern USA between AD 1130 and 1380, disrupting the Mill Creek Culture (Colorado). Krech (2001) suspected that ENSO events might be behind the decline of a number of North American Indian cultures, notably the Hohokam and O'odham, which flourished in the Medieval Warm Period (see later in this section). Australian Aboriginal peoples appear to have altered their 'tool culture' between 5000 and 6000 BP, possibly in response to climatic changes. However, it is not only climate change that might affect human cultures – they may also be damaged by very short-lived events like hurricanes, earthquakes and tsunamis, often leaving little record. In Florida the Calustra people sustained quite large settlements by hunter-gathering, but seem to have dispersed after a single hurricane or storm surge in the early eighteenth century.

Since the 1980s there has been more and more awareness of the desirability that human development is sustainable, that it seeks ongoing stewardship of resources, rather than short-lived exploitation, followed by degradation and hardship. Some past civilizations seem to have failed to maintain sustainable resource use; for example, the huge city and temple complex of Angkor Wat and Angkor Thom (Cambodia) flourished from the ninth to the fifteenth centuries AD; its builders, the Khmer people, then seem to have abandoned it. There are indications that the Khmer's exploitation of the region's forests and land caused environmental degradation that cut food supplies (Deighton, 2001). When cultures overexploit their environment, climate change, severe weather or sudden disasters like tsunamis or earthquakes have more impact.

The Roman Climatic Optimum and the Dark Ages

Between AD 400 and AD 1200, Europe became slightly warmer than now; some recognize a 'Little Optimum' between AD 750 and AD 1230; from AD 1200 to AD 1400 the climate became more variable. Archaeological and palaeoecological evidence suggests that upland regions like Dartmoor (southwestern UK) had been farmed in the Mesolithic (*c.* 4000 BC to 500 BC) only to be abandoned and later resettled between AD 800 and AD 1000; abandonment probably took place again in the late thirteenth and early fourteenth centuries AD.

Some scholars recognize a Roman Climatic Optimum between about 500 BC and AD 300. Over the few hundred years following it, the environment may have deteriorated. There have been suggestions that a comet or asteroid bombardment afflicted Europe around AD 540, causing climatic disruption and starting the Dark Ages. These claims are based on interpretations of legends and chronicles, and on a series of narrow tree rings that suddenly appear in the record for various parts of Europe, indicating a temperature fall (see Box 3.3). Alpine glaciers grew from about AD 450 to AD 700, so there may have been climatic deterioration around 540, even if a bombardment was not responsible. Whatever the cause, there seems to have been a blight on agriculture around AD 540, which may have debilitated people, making them more vulnerable to plagues around AD 542.

Bad weather and floods in China in the early 1330s AD have been linked to the outbreak and spread of bubonic plague; the blame for transmission has also been levelled at the movement of the Mongols from the fringes of the Gobi Desert between AD 1279 and 1350, a migration some argue was triggered by climatic deterioration. However, Keys (1999) suspected that Ethiopia and other parts of Africa were more likely to have been the source areas of plague

---

Box 3.3 European dendrochronology – 'narrow growth-ring events'

Growth lines traced from numerous wood samples, and often correlated with Greenland ice core data, have been interpreted to indicate adverse (cold or cool/wet) climate conditions at the following times.

AD 1741
AD 1325
AD 1177
AD 540
   207 BC
  1159 BC
  1628 BC
  2345 BC
  3195 BC
  4370 BC

Source: Various sources, including Baillie (in Slack, 1999: 55).

outbreaks toward the end of the Roman Empire. He explored complex chains of linkages between climate change and human fortunes, particularly the fall of the Roman Empire and the start of the Dark Ages. His thesis is that a drought and then heavy rainfalls caused a rodent 'population boom', and movement of gerbils and mice in rural Ethiopia and the Sudan. These creatures harbour bubonic plague (*Yersinia pestis*, widely held to have caused the Black Death), which they could have spread to the ivory and slave routes, along which the disease could travel to Egypt, eastern Africa and then to Byzantium, the Mediterranean countries and Europe where there were serious epidemics between AD 541 and 543 (the 'Plague of Justinian'). Periodic flare-ups took place until AD 750 and perhaps later. The plague-weakened Roman Empire failed to fight off aggression from tribes like the Vandals, Ostragoths and Visigoths, and went into final decline, with Byzantium conquered in AD 542. Climatic change might thus have *triggered* the collapse of the Byzantine Empire and other cultures, such as Athens in 430 BC.

Keys (1999) sought to reinstate respect for what he termed 'evolved determinism', tracing a series of human problems in the middle sixth century (including catastrophes in Greece, Byzantium, Rome, Mesopotamia, Persia, Korea, China, southwest USA and Central America) to environmental causes. He argued that these events, and others like the Saxon invasions of Britain and the rise of commerce and trade in northern Europe, may have been triggered by a planetesimal strike, or, more likely, the explosive eruption of a volcano. Keys identified the likely culprits: a powerful AD 535–536 eruption of Mount Rabul (or Rabaul), between Papua New Guinea and Java, or Mount Batuwara (Java/Sumatra). Similar disasters, he warned, will happen again, and he identified possible volcanic threats. Environmental historians propose numerous chains of causation; few are proven and seldom is there impassionate detective work to review all possible explanations and to sift the evidence.

There has been some controversy over the events around AD 536, when several observers recorded dry fog, dimmed sunlight and crop damage, yet so far there is no record in Greenland or Antarctic ice cores of the acid peaks that correlate with the explosive volcanic eruptions that might cause such phenomena. A similar widespread 'mystery' dry fog was recorded in AD 1783 by Benjamin Franklin in the USA and Gilbert White in England. Grattan and Pyatt (1999) investigated these and other dry fogs, which are reported to have dimmed the sun for lengthy periods, smelt of sulphur and damaged crops. They are inclined to blame volcanic gases in the lower atmosphere rather than the stratosphere. The effects vary from area to area due to weather circulation, with some fogs affecting large regions (another possible cause is discussed in Chapter 5: the marine outgassing of methane, sulphur dioxide and other gases).

It is widely accepted that volcanic eruptions can affect climate and, perhaps, indirectly trigger crop failure, epidemics, warfare, social and economic change. Mass and Portman (1989), and Sadler and Grattan

(1999) have investigated the dangers in linking disparate and geographically dispersed palaeoenvironmental data to known volcanic eruptions. These studies make it clear that several surface temperature declines that have been associated with eruptions were already under way well before the volcanoes were active. There might also be situations where eruptions are indirect causes of climate change; for example, ENSO events might be triggered by volcanic activity and then modify climate patterns.

Historical climatologists have correlated eruptions like Huanya Putina (AD 1600, Peru) with cool summers (that of 1601). There have been many attempts to link the Tambora eruption (AD 1815, Indonesia) and the poor summer in the Northern Hemisphere in 1816. That summer there were 15 centimetres of snow in New Haven (Connecticut, USA) on 10 June, and frosts during August – 'eighteen hundred and froze to death' as the year became known (Officer and Page, 1998). The eruption–chilling correlation is not always valid: meteorological records show summer frosts in Maine (USA) in 1778, 1794, 1797, 1800, 1796, 1835 and 1836, and no major eruptions are known just before these. So, frosts and snow in summer 1815 might not indicate a volcanic cause and have simply been occasional extreme conditions that just coincided. Similarly, the poor conditions in Scotland and England in 1783–84, which have been correlated with the Laki fissure eruption (AD 1782, Iceland) cannot be considered exceptional if one inspects extreme climate events on a timescale of a few centuries; they would be 'normal'. While Sadler and Grattan (1999) agree that volcanic activity can have a climatic effect it may not always be to blame, also it might cause cooling in one area and little change in others.

Critics of determinists like Keys argue that an environmental event that indisputably affects humans, and can accurately be pinned down, is relatively rare (for example, the eruption of Vesuvius in AD 79, which buried Pompeii and Herculaneum). Nevertheless, there have been some impressive pieces of environment–human impact 'detective work'. Medieval chroniclers in western Europe record strange outbreaks of an affliction they called 'St Anthony's fire', often suddenly striking virtually the whole population of villages or towns. The disease had distinctive and grotesque symptoms, a high rate of fatality and left many of the survivors with serious permanent debilitation. Modern epidemiologists have matched the symptoms with those of ergot (*Claviceps purpurea*) poisoning, caused by a blight (mould) infecting grain, especially rye. People in Europe tended to grow rye as a fall-back for when bad weather conditions hit normal grain harvests; if it is harvested and stored damp the mould flourishes. A few contaminated rye grains ground into flour are enough to decimate a village. Outbreaks of ergot poisoning coincide with other evidence of cool and wet weather, so help confirm climatic conditions.

Those investigating past environment–human relations can offer lessons to those hoping to predict future impacts of, say, global warming. Recently there have been assumptions that warming will cause an expansion of malaria transmission. However, a convincing study by Reiter (2000b), argues that malaria was rife in the UK and Europe *even throughout the Little Ice Age* (LIA), and that temperature changes had minimal effect on transmission. This is not just because some summers, even when the LIA winters were most chill, could be as hot as any today (e.g. 1661, 1665 and 1666); it may have more to do with housing, industrial pollution, agricultural and social changes (see Chapter 5). To assume that wetter or warmer conditions will encourage mosquitoes may be misguided. Mosquito larvae can be washed away by heavy rain and flourish in stagnant pools during dry conditions – simplistic assumptions are unwise.

The impact of similar successive climate variations on culture may differ because of environmental and social change and innovation. For example, from around the 1590s, but especially during the 1700s, northern Europe adopted the potato, which reduced dependence on cereal crops, and improved nutrition. The new crop was at first rejected in many areas (often for superstitious reasons), but gradually innovation took place and it probably helped offset some of the worst effects of the LIA. Ester Boserüp (1990) stressed that people faced with an environmental challenge or rising population are not restricted to making their own new inventions (in some stress situations that could be unlikely); they probably become more receptive to diffusion of new ideas from elsewhere, an ongoing process anyway. Where there are no barriers people might spread or move and continue in their old ways, but if those adaptions are constrained (especially likely with islands and with wetlands surrounded by desert) the task is more difficult; also, unreceptive people are more likely to fail to adapt to challenges.

The Medieval Warm Period

The Medieval Warm Period (MWP), Medieval Warm Epoch or Medieval Climatic Optimum, spanned roughly five centuries, and is seen by most environmental historians as a Golden Age for humans in Europe, North America and possibly elsewhere. The MWP is generally dated between AD 850 and AD 1300 (medieval record-keeping was to say the least, patchy – probably the bulk of the eleventh and twelfth centuries were warm). Some recognize a Little Climatic Optimum, roughly between AD 1050 and 1250 (Parry and Carter, 1998: 1). During this period, average mean temperatures (interpreted from a range of evidence) ranged from slightly cooler than now to *c.* 1.0 to 1.4°C warmer. The MWP may not have been fully manifest worldwide, but it does seem to have affected many Northern Hemisphere regions and some of the Southern Hemisphere at broadly the same times. There are records from China of the expansion of citrus trees, indicating January minimum temperatures as much as 3.5°C above present. The degree of warming probably differed from region to region, so the consequences were varied. Records of the siting of vineyards and palaeobotany suggest less frosty springs, warm summers and longer autumns before frost, especially between about AD 1100 and AD 1300.

Conditions during the MWP may have encouraged seafaring and deep-water fishing (the distribution of cod and herring shoals differed markedly in the MWP from that before and after), populations grew and villages multiplied. Vineyards flourished as much as 500 km north of mid-twentieth century limits, and records of grain harvests also suggest warm conditions. In central Europe until somewhere between AD 1300 and AD 1450 vineyards were cultivated at as much as 200 metres' higher altitude than today (Pearson, 1978: 59; Pfister *et al.*, 1999). In the seventh century AD, England had dozens of vineyards; some of these were still flourishing in the eleventh century (in AD 1085 the *Domesday Book* recorded 38, *in addition* to those owned by the King (Bryson and Murray, 1972: 68)). In the second half of the thirteenth century things seemed to have peaked, and by 1450 all English and Welsh winemaking had ceased. Only since the 1980s have vineyards been re-established in the UK. While the improvement of shipping, the loss of rural labour through plague, the destruction of the monastic system and the rise of a merchant class in Europe may have played a part, the likelihood is that cooler and wetter conditions were to blame.

Glaciers in the Alps receded so much during the MWP that, in Medieval times, it became possible in some regions to re-open prehistoric copper mines (Le Roy Ladurie, 1972). In Europe the twelfth and thirteenth centuries were a time of population growth, settlement of new land (some of this in uplands was abandoned when conditions deteriorated), exploration and trade, and especially active church and (Gothic) cathedral building (Fagan, 2000: 19).

Archaeologists and environmental historians have examined the expansion of Norse settlements during the MWP, and their misfortunes in the subsequent Little Ice Age (see Box 3.1), many interpreting their changing fortunes as examples of environmental impact on human affairs. By the mid-seventh century AD the Norse (and some other Celtic peoples) were trading with Europe, Africa and Mediterranean countries. During the ninth century, Irish Anchorites and Norse settlers reached Iceland; slightly later, the Norse colonized Greenland and began visiting North America. European-type artefacts and turf dwellings suggest ship repair and temporary stays at L'Anse aux Meadows (northern Newfoundland), possibly parties gathering timber to take back to Greenland (Dansgaard *et al.*, 1975). The *Grœlendinga Saga* (sagas are epic tales) and *Eirik's Saga* make frequent mention of visits to 'Vinland'; this could have been Labrador, Newfoundland, or – given the descriptions of Indians and temperate vegetation – possibly even the US mainland as far south as New England. Wheat was grown in Norway to at least 52° 5' north (some claim to 64° 0' north) between AD 1150 and 1300. This seems to support the argument that Norse settlers enjoyed conditions warmer than those since the fifteenth century.

In Equatorial East Africa the MWP seems to have been characterized by conditions that were drier than now – especially between AD 1000 and 1270. That part of Africa became wetter during the Little Ice Age, although this wet phase was punctuated by at least three prolonged dry periods that may have been worse than any droughts experienced during the twentieth century (Oldfield, 2000). California had warmer and possibly wetter conditions than now during the MWP, and southwestern USA had higher lake levels between AD 700 and 1350. There is evidence during the MWP of glacial retreat in Europe, the USA, South America, New Zealand, Alaska and elsewhere.

The Little Ice Age

The MWP was not to last; cooling seems to have been under way in Greenland and the Arctic around AD 1200. In Europe, by the mid-fourteenth century, conditions had deteriorated; especially cool and wet summers were chronicled in 1315 and 1316 (Tuchman, 1978). The latter raised wheat prices as much as eightfold and hit salt production (without sunlight and dry brushwood output plummeted). This in turn meant that people faced increased costs and may have had insufficient supplies to preserve adequate quantities of meat for winter; in those days a lot of livestock had to be slaughtered in the autumn because there was insufficient fodder to last until spring. These events may have caused the severe famines that ravaged western Europe, notably the 'Great Hunger' (or Medieval Great Famine) of 1315–19 or 1322, although, undoubtedly, socio-economic causes were also at play (Burroughs, 1997: 26). There were more serious European famines in the 1420s and 1430s, perhaps triggered by unfavourable weather. These caused economic depression and social unrest, widespread malnutrition and the movement of refugees, soon followed by outbreaks of bubonic plague.

Between roughly AD 1300 and 1850 (dating varies a lot, with the extremes for start and finish being AD 1250 and 1910; a reasonable range of dates is 1450–1890, a more conservative range is 1550–1750) average conditions, at least in the Northern Hemisphere, became cooler than they are at present (Grove, 1988). The term Little Ice Age (LIA) was originally applied to the most recent major glacial advance of the Holocene; it is now associated more with the advance of European glaciers between about AD 1450 and 1850 (Ogilvie and Jónsson, 2001). The LIA, according to Fagan (2000: xii), had a subtle but vast effect on human development for over 500 years, reminding us that natural climate change is inevitable, unpredictable and sometimes vicious. Fagan described the LIA as a '500 year-long cold snap in which thousands of European peasants starved' (1999: 181).

The LIA cooling was probably global: valley glaciers in the northern and southern high latitudes and in tropical highlands expanded, including those in South America, Africa and Papua New Guinea. Le Roy Ladurie (1972: 226) noted that European glaciers seem to have grown after roughly AD 1590, reached a maximum about 1670, and have been in retreat worldwide since about AD 1750 (before human activity really started to cause global warming). However, there was no significant high-latitude ice sheet glaciation – conditions were essentially those of a cold stadial phase, not a full glacial (Kaser, 1999). This cool phase may have been felt at slightly different times around the world, and was interrupted by warm spells and certainly some hot summers (Pearson, 1978: 221). Evidence around the Atlantic shows some synchrony, and records in China note the loss of long-established orange groves in Jiang-Xi Province. The LIA seems to have been at its worst between roughly AD 1550 and AD 1750 (in the UK possibly AD 1680–1730) (see http://www.catalinaconservancy.org/weather/lia.htm, accessed 25 May 2000).

The indications are that weather patterns shifted, altering wind tracks and rainfall over northern Europe to cause the LIA. Changes in the circulation of the Atlantic, possibly changes to the NAO or altered flow of the Gulf Stream might have been involved (see Chapter 5). The LIA also coincides with unusually low sunspot activity (a very low phase – the Maunder Minimum – between 1645 and 1745), but this may be coincidental. Researchers have also observed the recurrence of major volcanic eruptions during the LIA, notably El Chichon (Mexico, AD 1259), Oraefajokull (Iceland, AD 1362), Mount St Helens (USA, AD 1479) and as mentioned earlier, perhaps to blame for the terrible summer conditions in North America and Europe in 1816, Tambora (Sumbawa Island, Indonesia, AD 1815). There were at least two other large eruptions within a few years of Tambora: Soufriere (St Vincent, AD 1812) and Mayon (the Philippines, AD 1814).

The LIA was a period with some very harsh winters, and often cool and wet summers, but not all summers or winters were colder than now. It was essentially a period of changeable and unpredictable climate. Average temperatures were roughly between 0.5° and 2°C below those of today; precipitation fell markedly below that of preceding centuries and there was more variability in seasonal weather. There are records of sea ice floating to lower latitudes, and snowfall took place at lower latitudes and elevations in both the Northern and Southern Hemispheres. The Baltic and other European waterways froze on a number of occasions in the sixteenth and seventeenth centuries. There are also records of sea ice off Iceland and around the coasts of Britain and Europe. At the peak of the last glacial (*c.* 18,000 BP) permanent snowlines in European highlands fell roughly 350 metres below present-day levels, during the LIA they fell about 100 metres.

Some countries have records that indicate how people adapted to harsher conditions; in the European Alps and parts of northern Europe tax exemptions and reductions reflect hard times as the growing season fell by as much as 20 per cent. It has been suggested that the LIA helped encourage European settlement of the tropics, and while Greenland, Iceland and Scandinavia withdrew from seafaring and trade, countries further south in Europe (the Basques, Portuguese, English and Dutch) seem to have become more adventurous. Officer and Page (1998: 100) went so far as to trace the origin of present-day troubles in Ulster (Northern Ireland) to the Scottish migrants who were encouraged to settle there by the LIA conditions during the reign of James VI of Scotland (in the seventeenth century); conditions they also suggest may have helped force the Scots into union with England. Between AD 1693 and 1700 Scotland suffered seven out of eight years of failed harvests. Direct climatic causation was probably not the only factor at work – during the three or four centuries before the LIA there had been population growth and agricultural expansion in Europe and elsewhere, some of this may not have been sustainable even if the climate had stayed benign, and soil degradation, deforestation, overgrazing and excessive game hunting had probably already started to bite. In England the population is estimated to have increased from around 1.4 million at the start of the eleventh century to roughly 5 million at the start of the thirteenth. Today the population of the UK is more than ten times that of 1300, and one wonders what effect a modern LIA might have. Debilitation of people over a few successive years of poor harvests caused by cool and wet weather events probably led to reduced labour efficiency and availability (due to increased deaths), which hit further harvests. A spiral of debilitation may have paved the way for the spread of epidemic disease, notably plague, but also smallpox and many other infections that probably helped cause abandonment of some settlements.

In the UK, and elsewhere, agriculture tends to expand into less favourable areas, like uplands, when conditions improve, especially when there is high demand for produce, only to contract from these marginal environments when conditions or demand deteriorate. A number of 'lost villages' in the UK may be the result of climatic deterioration or sea-level changes depriving them of harbours or silting up channels. Higher altitudes, more exposed locations and areas of poor soil are especially prone to abandonment when climate deteriorates. However, some settlements may have collapsed through disease or the actions of ruthless landlords (Beresford, 1954; Beresford and Hurst, 1971). In Eire in the mid- to late 1840s crop disease was clearly a cause of depopulation; but the people had become dependent on potatoes for social, economic and political reasons.

The LIA is more accessible than earlier climatic changes. It is a period for which there are some written records and even a little meteorological data, as well as palaeoecological evidence (Lamb, 1977; Grove, 1988; Fagan, 2000). It is the youngest of three relatively long cold spells since the last glacial. Knowledge about the LIA has been improved by ice cores from Greenland, Antarctica, the Rockies and the Andes. However, reliably linking the LIA with human affairs is not easy.

While caution is needed before attributing blame for human problems to climate change, opinion can also swing too far the other way; Pearson (1978: 228) noted that 'The effects of the (*climatic*) deterioration from 1300 to the sixteenth century have . . . commonly been overlooked because of the overwhelming concern with the Black Death.' This is not surprising: between 1346 and 1400 bubonic plague wiped out around one-third of Europe's population (see Chapter 5).

Even apparently unfavourable conditions like the LIA may have beneficial impacts on humans; the possibility that it helped stimulate trade and
exploration by southern European nations has been discussed. It is also
conceivable that adverse conditions encouraged innovation that might
otherwise not have happened before population and land degradation made
it impossible. Fagan (2000: 107) argued that the LIA might have played a
catalytic role in the modernization of agriculture and land reclamation in
Britain and countries like Holland; perhaps it even helped prompt the social
change and industrialization that led to modern Europe.

Recent studies of ice cores from Greenland, Antarctica and elsewhere
show that the end of the LIA was rapid, taking no more than ten years and
probably only three to four. This serves as a warning: if change took place
quickly in the past, then it could do so in the future, even without the uncertainty of greenhouse warming.

The Modern Warm Period

The period following the LIA, but especially from *c.* AD 1850 to the present,
has been termed the Modern Warm Period. A number of historians and climatologists have noted that the twentieth century has been unusually favourable
compared with earlier environmental conditions. However, during this period
there has been occasional harsh weather, especially the winters of 1740, 1929,
1947 and, notably, 1963. There have also been markedly cool summers: 1725,
1740, and, in particular, 1816. In 1963 UK agriculture suffered badly: around
20 per cent of sheep died, and the severe weather had significant economic and
political impacts, including helping to bring down the UK government of the
day (the situation was similar in 1947) (Burroughs, 1997: 61). It is still not fully
clear just how much adverse weather was to blame for the Irish Potato Famine
(1844–46; some date it 1845–49); the fungal infection responsible for rotting
the crop, *Phtopthora infestans*, may well have been aided by unusually wet and
mild winter conditions. Following an outbreak in the USA in 1843 most of
Europe was infected by 1845; in Ireland the poor were more dependent on
potatoes and the debilitated victims were decimated by diseases, especially
outbreaks of typhus in 1847 and cholera in 1848. Of around 8 million people
in Southern Ireland (Eire) 2 million had died or emigrated by 1851 (Couper-
Johnston, 2000: 79). The famine may well have generated 250,000 migrants,
so far Europe's greatest recorded eco-refugee exodus.

The UK and Europe, other than through disruption caused by warfare,
has had no serious hunger since the 1840s. The improvement seems to have
been achieved through agricultural development, the construction of a
railway network, and steam shipping. Burroughs (1997: 44) suggested that
it was not until the 1880s that UK technology really reduced vulnerability to
climate change. Even if technology has given richer humans a new advantage since the 1850s it would be wise to note that:

Through the millennia of human history, we have matched our rhythms to those of the Earth. But recent evidence from ice cores and elsewhere has shown that we have developed agriculture and industry during an unusual period of stable climate.

(Albey, 2000: 20–1)

Hunger is still a threat in many poor countries, and population growth plus environmental degradation are eroding the improvements that have been made.

During the Modern Warm Period there has been wider settlement than ever before, including large swathes of Australia, North America, New Zealand and the savannahs of the New World and Africa. Land development, vegetational changes and growing industrial pollution are causing anthropogenic climate changes. Added to the natural uncertainties of the past we now have environmental shifts caused by humans – change that is difficult to predict. Were it not for these anthropogenic effects it would be a reasonable guess that the present day is at roughly the midpoint of an interglacial. Will anthropogenic changes counter unfavourable natural changes, or prompt unwanted, perhaps sudden, environmental problems?

3.2 El Niño, La Niña – El Niño Southern Oscillation (ENSO) and the North Atlantic Oscillation (NAO)

El Niño

El Niño, 'the Christmas (male) child' is a recurrent event named by the Spanish colonists in Peru in the sixteenth century. It is an ongoing climatic variation (see Chapter 5) marked by the arrival of warm water, often in late December or early January, along the Pacific coast of Latin America, which at other times is washed by the cold Humboldt Current (see also Chapter 5). El Niño events are the warm extreme of an El Niño Southern Oscillation (ENSO) continuum, an ongoing ocean–atmosphere oscillation; the other (cold) extreme of the ENSO are the La Niña events. El Niños are not always clearly defined and discrete, some can be very unpredictable (see Box 3.4). Successive El Niños can develop in quite different ways, and so forecasts must be treated with caution. However, their general progression is known, and can allow some predictions (Couper-Johnston, 2000).

El Niños manifest off southwest Ecuador and northwestern Peru when warm water spreads eastward across the Pacific together with Equatorial lower-pressure cells. The effect is to cause Equatorial Pacific precipitation to move eastward from Indonesia and a catastrophic reduction in the upwelling of nutrient-rich cool water along the western coast of South America (Fig. 3.3). Consequently, huge shoals of anchoveta (*Engraulis ringens*), which feed off the west coast of South America, starve because El Niños decimate plankton; there may also be blooms of poisonous algae

Box 3.4 Some El Niños and La Niñas

Year (AD)	Coincidental – perhaps related – events
550	Drought in Amazonia
1000	Drought in Amazonia
1200–01	Drought in Amazonia
1396–1410	
1500	Drought in Amazonia
1595–96	
1629–33	
1640–41	
1685–88	Strong El Niño
1780	*La Niña event* – hurricanes in Atlantic
1788–93	Series of linked El Niño events
1812	Severe winter Russia
1814	
1817	
1876–78	Famine in China, drought northeast Brazil/India
1884	
1891	
1904–05	
1911–15	An especially long event, floods in China
1917	
1918–19	
1925–26	Drought in Amazonia
1930–1	
1935–36	Drought in Midwest USA
1940–42	Severe winter in Europe
1950	*La Niña event*
1957–58	
1965	
1972–73	USSR, large grain purchases from abroad after poor harvest
1974	*La Niña event*
1972–73	
1976–77	
1982–83	Especially strong forest fires in Indonesia
1987	
1988–89	*La Niña event*, famine in Ethiopia
1993–94	Bushfires in Australia
1990–95	An especially long event
1997	Severe El Niño, drought in India and Southeast Asia
1997–98	One of the strongest events – forest fires in Mexico – worldwide it is estimated to have cost US$32 billion and caused some countries to be crippled economically
1998–99	*La Niña event*

Source: Compiled by author from various sources.

Notes

This is not a complete listing.

Most El Niño events last about 12 months and take place roughly three to five years apart.

There have been suggestions the severe El Niño events of 1982–83 and 1990–95 were triggered by volcanic eruptions: El Chichon (Mexico) 1981, Mount Pinatubo (Philippines) 1991 and Mount Spurr (Alaska) in 1992.

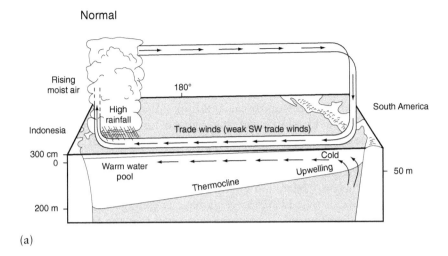

(a)

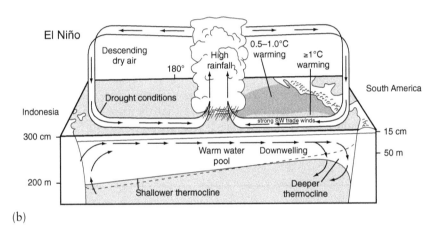

(b)

Figure 3.3 ENSO: (a) normal conditions, and (b) El Niño taking place
ENSO is essentially an oscillation between high air pressure over the southeast
Pacific and low air pressure over Indonesia. In (a) – normal conditions – air
pressure differential drives easterly trade winds along the Equator. Warm water is
blown into a 'hump' in the eastern Pacific and this depresses the thermocline there.
Off the west coast of South America, cool water wells up, supporting plankton and
rich wildlife, and reducing the amount of rain reaching the shore. Indonesia gets
moist westerlies and the eastern Pacific is dry. In (b), the east–west air pressure
differential is reduced and trade winds (westerlies) die down. Warm water flows
eastward near the surface. An upwelling of cold water off the west coast of South
America declines, with catastrophic effect on wildlife. Onshore winds bring more
rain to western South America. This is an El Niño.
Source: Redrawn, with modifications, from various sources.

(red-tides or *aguaje*), which render fish and shellfish unfit to eat. For Peru and adjoining countries El Niños mean reduced seafish catches, damage to wildlife, rainfall that disrupts irrigation, and floods that wreck towns and sweep away infrastructure. At higher altitudes in the Altiplano of the Andes El Niños can trigger short but severe droughts.

El Niño events have wider impacts, probably almost global (through a chain of tele-effects), affecting Africa, Amazonia, Australasia, Asia and high latitudes. The 1998 El Niño event was blamed for weather problems in California, Mexico, Brazil, and Southeast Asia, notably for dry conditions that led to forest fires in Indonesia. Naturalists suggest that coral reef damage and synchronous flowering of tropical rainforest trees in Southeast Asia and elsewhere are related to El Niños.

Recent studies of past El Niño events have encouraged a form of neo-environmental determinism. It has been suggested that the fall of the Old Kingdom of Ancient Egypt in 2180 BC was triggered by drought, which may well have coincided with an El Niño event. Fagan (1999: xiv) noted that,

> Until recently, scientists studying ancient civilizations and those specializing in El Niño rarely spoke to one another. Now they work closely together, for they realize that this once-obscure Peruvian countercurrent is a small part of an enormous global climatic system that has affected humans in every corner of the world.

In the past droughts, floods and temperature variation were seen as random vicissitudes, there is now growing suspicion that they are largely recurrent events and might be modelled and predicted (see Box 3.5 and Fig. 3.5). It is widely accepted that when a strong El Niño affects Peruvian coastlands

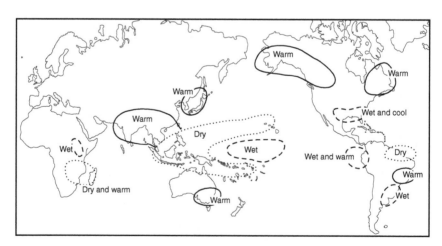

Figure 3.4 Land–ocean–atmosphere interactions: the thermo-haline conveyor (flow patterns much simplified)
Source: Redrawn from various sources

Box 3.5 Land–ocean–atmosphere interactions – the salo-haline conveyor

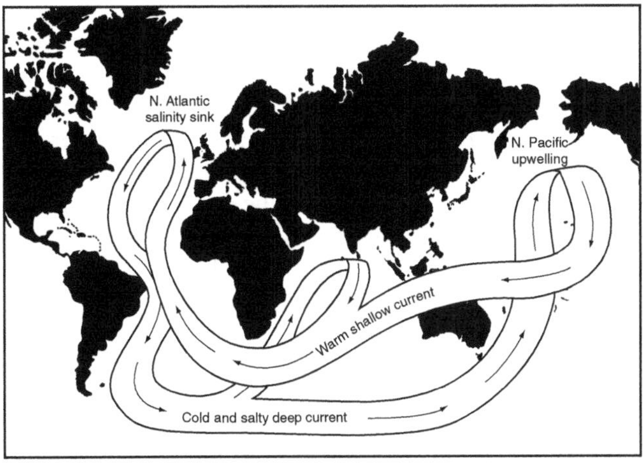

Figure 3.5 El Niño impacts (October to March)
Note: Drought in southern Africa, northern Australia and Indonesia; high rainfall in western South America, southeastern North America, the western Pacific and eastern Africa (see also Fig. 3.4b).
Source: Redrawn from several sources.

The Ocean-Global Conveyor (or Great Ocean Conveyor) takes the form of cold and salty deep water flows from the North Atlantic southward and then into the Pacific, where it upwells and returns as a shallow, warmer and less salty flow. During cold climatic phases (glacials) the Conveyor seems to be interrupted. Gradually, information on ocean convection cells is accumulating. For water to circulate once around the Conveyor probably takes 500 to 2000 years – so the system now includes water affected by the climate of centuries ago. Factors including the winds, coriolis force, convection, and especially salinity and temperature differences, determine the circulation. Warm surface water flows towards Greenland and the Arctic where it cools (becomes more salty and denser), and sinks and flows towards the southern Atlantic and Antarctica at depth. The scale is huge.

The Gulf Stream is a huge eddy (gyre) driven by winds and the Earth's rotation, and possibly affected by changes in the Conveyor; a 'turning off' of the Gulf Stream would chill western European winters.

In the past, ice melt in North America and Greenland may have upset North Atlantic circulation by flooding less dense freshwater into the sea; at times (possibly 8000 to 9000 BP) the disruption was sufficient to 'flip' the system into a mode with the warm Gulf Stream shifted southward. This southward shift would mean sudden and severe cooling for the UK, Iceland and western Europe and parts of the USA, especially in winter. Today some researchers are worried that global warming might trigger similar shifts as Greenland's ice caps and Arctic ice sheets melt.

Melting ice flowing into the North Atlantic could alter salinity and affect circulation (this may have happened in the past, including around 8400 years BP). If the Gulf Stream is weakened, Europe could suffer a 10°C fall in winter temperatures.

Sources: Text based on various sources, including Alley, 2000; www.epa.gov/global warming/glossary.html (accessed December 2001); Fig. 3.5 drawn by the author from various sources, including Burroughs, 1997: 129, Fig. 5.15.

there is a good chance of drought in northeast Brazil and very dry conditions in Southeast Asia some months later. An El Niño has even been blamed for the unusually southerly icebergs that destroyed the *Titanic*; another may have increased the world price of fishmeal, sent soya prices rising, and helped encourage the use of cattle and sheep carcasses in animal feed, leading to the bovine spongiform ecephalopathy (BSE) problem in the UK. However, yet again, *proving* such causation is another matter.

Many peoples seem to have adapted to 'boom-and-bust' food availability, which may be related to El Niño events. The adaptions, seen in Papua New Guinea, northwestern North America and Australia, take the form of seeking alternative food sources and relocating well away from normal habitats when 'bust' conditions strike. Couper-Johnston (2000: 55–96) suggested that this could be the reason for ceremonies like 'potlatch' in Canada and Alaska, and for Aboriginal 'songlines' and 'walkabout' in Australia. Societies may have developed El Niño-related reciprocity arrangements: typically feasting and gift-giving in 'boom' years to cement links with people who can be called upon in lean years. 'It is perhaps significant that the vast majority of hunter-gatherer communities that have survived into present times exist in zones of El Niño influence' (Couper-Johnston, 2000: 75).

LA NIÑA

The '(female) child' is essentially the cold opposite of an El Niño. During La Niña events warm water appears in the western Pacific (and cold surface water in the central and eastern Pacific), leading to increased convection and rainfall over western Latin America. These events sometimes cause increased rainfall in desert areas of Ecuador and Peru, and have been correlated with flooding in the Sudan, Bangladesh and Thailand in 1988. It seems they also impact on conditions in China, Brazil, Indonesia and Australia (where Lake Eyre may be replenished during La Niña events), and elsewhere. In 1998–99 severe hurricanes in the Atlantic were linked to a La Niña event; subsequently Couper-Johnston (2000: 47) reported that 22 out of the 26 most severe hurricanes in the past 100 years coincided with La Niñas. La Niñas seem to have less effect on Pacific storms, but what is not clear is whether La Niñas somehow increase Atlantic hurricanes or El Niños suppress them. Less is known about La Niña events than El Niños; it is likely they too have very marked impacts on many parts of the world (Glantz, 2002).

ENSO

El Niño and La Niña events do not show a regular periodicity, they vary in timing and strength and can unfold in different ways; they are the manifestation of an unstable ocean–atmospheric interaction that 'flip-flops' back and forward. While there do seem to be wide impacts within the tropics, and probably also at higher latitudes, it is not clear whether they have global effect (such as raising global temperatures). Recent reports have suggested that since the beginning of the last century, ENSO events have been stronger than in the previous 130,000 years (based on fossil coral reef evidence from northern Australia and Papua New Guinea) (Grove and Chappell, 2000: 7–10, 35).

There is evidence that the effects of, at least the more marked, ENSO events are felt in southern USA (including the Rockies and possibly the Great Plains); in Indonesia, India, Kenya, South and Southeast Africa, Ethiopia, the Sahel, the Caribbean and northeast Brazil, some droughts have been correlated with it. Even when there is no direct impact, world food prices may be driven up during ENSOs. It looks likely that the highlands of Papua New Guinea and highlands in Irian Jaya, mainly above 2200 m altitude (but occasionally down to 1450 m), suffer frosts during El Niño events, and at lower altitudes droughts seem to parallel the frosts. These frosts and droughts developed in 1902, 1914, 1941, 1972, 1982 and 1997–98 (Grove and Chappell, 2000: 113). Late 1990s droughts led to serious forest fires in Kalimantan and Sumatra, in turn leading to widespread smoke pollution across a wide belt of Southeast Asia. The combination of El Niño drought plus forest clearance by settlers seems to make even moist forest areas vulnerable to fires. There are also signs that flood and drought along the Nile, in central India and in the Darling River basin (Australia) take place during ENSO events. A number of disease outbreaks coincide with ENSO events, notably the 1992–93 Hanta virus infections in southwestern USA, and the 1845 potato blight in Ireland (see Chapter 5) (Grove and Chappell, 2000: 89–108; Caviedes, 2001). Correlations seem to be highly significant, although not proven (and the link is not always evident), forecasting can be attempted (see Fig. 3.6), perhaps giving several months' warning, but caution is needed (Grove and Chappell, 2000: 12; Whitaker *et al.*, 2001).

NORTH ATLANTIC OSCILLATION

There is a less well understood northern parallel of ENSO: the North Atlantic Oscillation (NAO) (Fig. 3.7). The NAO is a constantly changing pressure gradient over the North Atlantic, an ocean–atmospheric interaction like ENSO. A predominantly wintertime phenomenon that influences the conditions in the North Atlantic, UK and Europe, it usually manifests as a high-pressure cell over the Azores. The NAO plays a major role in helping determine Europe's winter rainfall and storms, and to a lesser extent summer weather (for further details and references, see Matthews, 2001: 432). The workings of the NAO ocean–atmosphere interaction are not fully understood; however, during the LIA a low-pressure cell appears to have been present near Iceland (Fagan, 2000: 23–6). The NAO oscillation might at one extreme help cause cool conditions like the LIA, and at the other warmer conditions like the Medieval Warm Period and Modern Warm Period (see discussion of the variation of the Gulf Stream and the thermohaline conveyor in Box 3.5 and Chapter 5). When high pressure develops

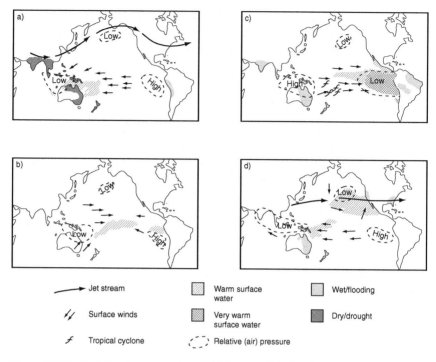

Figure 3.6 Typical development of an ENSO event (idealized sequences)
Notes
(a) Normal (January)
(b) Pre-ENSO (September)
(c) ENSO event (December)
(d) Post-ENSO (January)
Source: Reproduced with permission from Barrow, C.J. (1991) *Land Degradation Development and Breakdown of Terrestrial Environments.* Cambridge University Press.

over the Azores (a high NAO index) the Atlantic off the east coast of the USA can warm (typically in September); northern Europe is then likely to experience winter storms and flooding, and the Mediterranean and parts of Asia tend to have drought. A high NAO index developed between 1900 and 1930. A low NOA index is likely to mean summer droughts in northern Europe and more severe water shortages in the Mediterranean. There is a possibility that global warming will trigger a high NAO index.

3.3 Extreme weather events

Extreme weather events occur even in the most benign periods and will continue to happen in the future, greenhouse warming or no greenhouse warming; whether the frequency and severity alters remains to be seen.

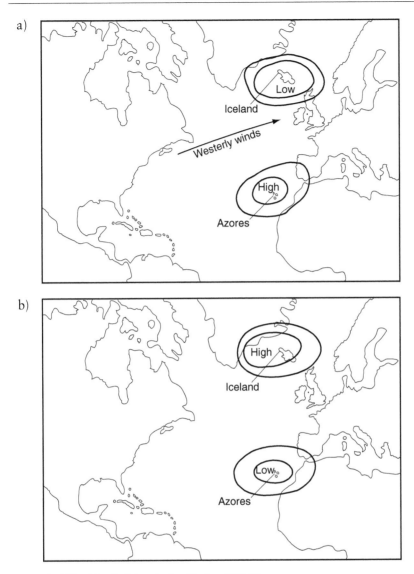

Figure 3.7 The North Atlantic Oscillation (NAO)
Notes
(a) High NAO index (an NAO 'event') – low air pressure around Iceland, high
over Portugal and the Azores, resulting in prevailing westerly winds that warm
Europe (and bring storms). With the Gulf Stream flowing, this keeps Europe warm
in winter and improves rains in summer – the present-day situation.
(b) Low NAO index – reduced pressure gradient results in weakened westerlies.
Europe is markedly cooler and air flows from the Arctic and Siberia are more likely
to cross Europe. The oscillation from (a) to (b) can be sudden and unpredictable;
sometimes there is little change for decades. Alterations of the Conveyor (see Fig
3.5) might also affect the Gulf Stream.
Source: Redrawn from several sources.

Extreme events may be relatively brief but they can still markedly affect human affairs (see Box 3.6 and Fig. 3.8). These events are often greeted with 'surprise' but they happen even when conditions are 'unchanging' and 'stable' (see Section 5.9, on severe weather events, in Chapter 5). Burroughs (1997: 91) captured the widespread attitude of governments toward extreme events like severe winters, droughts or floods thus: 'short-term panic measures and long-term complacency'.

Warfare has been one human activity much influenced by extreme weather (Burroughs, 1997: 54–7). For example:

- Kublai Khan's invasion of Japan in AD 1281 failed due to storm winds
- the Spanish Armada of AD 1588 was probably defeated as much by August gales as by the English fleet
- Napoleon I, in 1812, may have fared better with his invasion of Russia were it not for unusually severe winter conditions
- In 1889, German, British and US ships grouping near Western Samoa were decimated by a hurricane; war did not materialize – instead the Treaty of Berlin was signed.

Box 3.6 Selected extreme weather events recorded over the last 500 years

- The Great All Souls Day Storm, 1–6 November 1570
- Heavy snows in England, 1579
- Severe frosts in Europe, 1608
- Heavy snow in England, 1615
- Severe frosts in Europe, 1681
- Harsh frost in England and extensive sea ice around UK and European Channel coasts, 1683–84
- Great storm strikes UK, 1704 and northern Europe – may have killed 400,000 (a huge number given the population of the time)
- Unseasonable (mid-summer) frost and snow (and a cold winter) in Europe and USA, 1815–16
- Severe cold and snowfalls UK and Europe, 1947
- Storm and tidal surge floods large areas of The Netherlands and UK, 1953
- Severe cold and snowfalls UK and Europe, 1962–63
- Widespread snowfalls in South and southwest Africa, 1964
- Drought in eastern USA, worst since records began (1738), 1963–64
- Baltic Sea completely frozen, 1965–66
- Sahelian drought, 1968–73
- Sea ice stops shipping to Iceland for first time since 1888, 1968
- Coldest winter on record in eastern Russia and Turkey, 1971–1972
- Great storm (southern UK), 1987
- Hurricane Gilbert (Caribbean), 1988
- Hurricane Hugo (Caribbean), 1989
- Hurricane Andrew (Caribbean), 1992
- Blizzard in eastern USA, 1993

Figure 3.8 Teotihuacan (Mexico) 'pyramid of the sun', *c.* 215 feet (72 m) high
This culture was at its peak around AD 650. The surrounding settlement of
between 125,000 and 250,000 was probably abandoned at the same time as other
parts of the world suffered climate-related disruption starting around AD535–39.
Source: Author 1999.

- cold conditions in 1940, 1941 and 1942 probably helped hinder and
  delay German attacks on the UK, giving the RAF time to add something
  like 20 squadrons before the crucial Battle of Britain; severe winter
  conditions in Russia during 1941/42 cost the Germans 110,000
  fatalities (attributable to exposure), and helped ensure the campaign
  was a failure
- in 1944, the US Pacific fleet was hit by a typhoon, which delayed invasion
  of the Philippines
- the 1944/45 Ardennes Offensive may have made less progress against the
  Germans if conditions had not been so cold.

FLOODS

Floods can be generated by a range of single or combined factors: single or
multiple storms; extended periods of steady rain; high tides and tidal surges;
spring melt; tsunamis; the damming of rivers by landslides, debris or ice,
and then sudden release; land use changes (channelization of rivers, installation of land drains, removal of vegetation and compaction of watersheds);
a storm over an already saturated area. The incidence and impact of
flooding can be affected by human activity and social development. Land

clearance can increase run-off, causing flooding that would not happen if the vegetation were undisturbed. The drainage of agricultural land and riverine wetlands, and development of floodplains reduces the areas available to act as flood buffer zones, and exposes more people and property to damage. Channels and culverts that cope with floods for years may fail if people fly-tip rubbish in them or neglect to maintain them. Altered farming practices may increase riverine floods: in the UK, the installation of field drains and the practice of planting winter cereals on flat-harrowed fields rather than along contour furrowed fallows (which trap and hold water better) increase and speed up run-off.

River valleys have been the stage for much human history, and many peoples have adapted to river regimes and successfully made use of the available riverine resources. There are large gaps in the archaeology of the Congo, Amazon and other tropical floodplains, which may have supported flourishing cultures when higher latitudes were in the grip of the last glacial. Some of the earliest civilizations were essentially 'riverine': those of the Tigris-Euphrates, those along the Indus and Egyptian settlements beside the Nile. Other settlements along rivers have proved amongst the most successful and sustainable. Many people have traditional building styles that seek to minimize flood risk – for example, homes built on high wooden piles are common in Southeast Asia. Although similar dwellings were used in prehistoric times in Europe, the practice has strangely not been continued in flood-prone regions like The Netherlands. Not all riverine settlements are flood-adapted, even when the risks are known.

STORMS

Storms affect human fortunes worldwide, from the high latitudes to the tropics. Storms may occasionally shape history by influencing the outcome of warfare, but it is generally difficult for historians to pin the decline of a society on a single extreme storm event; although there are legends of townships, including some along the North Sea coast of the UK, being destroyed. Archaeologists have suggested that the Calusa people, a hunter-gatherer group that developed townships in coastal southwest Florida, may have had their dwellings and social organization destroyed by a hurricane around AD 1710. (Hurricanes are discussed in Chapter 5 as an ongoing threat.)

DROUGHTS

There are accounts of what are probably drought-related famines in the Old Testament (around roughly 3500 BP). Humans have always been affected by drought, although that vulnerability increased when hunter-gatherers became agriculturalists tied more to a single location. Archaeologists and environmental historians have turned to drought to explain the demise of past civilizations, often with limited proof. The Egyptian civilizations carefully monitored Nile flows, knowing that weak flooding meant poor harvests and probably civil unrest. Throughout South Asia sophisticated tank rainwater storage and irrigation systems were developed to combat annual variations in monsoon and post-monsoon rainfall in order to try and ensure adequate food supplies. The Andes of Latin America get their name from the widespread terracing at least partly used to conserve soil moisture or to practise irrigation.

Huge sums of money have been invested in irrigation since the 1940s; it is the sector where considerable improvements of agricultural yield have taken place. The whole purpose of irrigation is to supplement natural precipitation, sometimes to boost or diversify yields, but often to improve security against drought. In the past, and still today, irrigation can fail through poor design inadequate management, gradual salt accumulation, social change, or sudden storms and earthquakes.

There has long been debate about the decline of agriculture in northern Africa since Roman times; the tendency has been to blame the drying climate. But it is more likely to reflect altered trade, reduced availability of cheap slave labour, breakdown of social capital (resulting in neglect of communal irrigation) and possibly land degradation as a result of exploitative farming. Studies of long-disappeared agriculture in the Negev Desert, Israel (Nabatean civilization, 300 BC–AD 650), have established that their run-off agriculture fell into disuse for social or economic reasons, not climatic. Indeed, basic refurbishment of ancient field systems has revealed that they still work with present-day rainfalls (Barrow, 1999: 49–73, 91–103).

Vulnerability has not been much reduced for modern rain-fed farming: drought and hardship struck Midwest USA in the 1930s (Box 3.7 and Fig. 3.9), the USSR in the 1950s, 1960s and 1970s, and the Sahel between 1968 and 1973 (in some Sahelian regions it persisted until the 1990s). Ethiopia and the Sudan were hit in the 1980s, southern Africa in 1991/92, and at regular intervals through the last few centuries northeast Brazil and many other countries have suffered. A significant proportion of today's traded grain is grown in areas subject to marked rainfall fluctuation and sometimes severe drought. There has been huge investment in irrigation, but as a means of controlling nature it not always successful.

Drought is not only an agricultural problem: as cities grow and tourism takes people to areas with limited water supplies, especially islands, demand for water increases. Worldwide, those people able to afford it are making ever greater demands for water, competing against agriculture and other users. California is facing something of a water supply crisis; around the world large and fast-growing cities are seeking to expand their supplies; even the UK (seen as a 'wet' country) ran short of water in 1976.

Since the 1970s there has been debate about desertification (the degradation of land to something similar to a desert); in the past this was generally

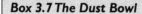

Box 3.7 The Dust Bowl

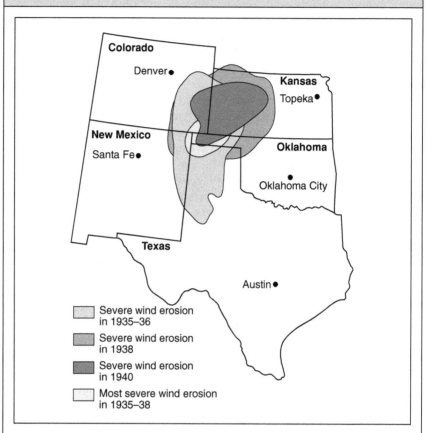

Figure 3.9 Location of the US Dust Bowl

The Midwest of the USA, which for the last century or so has included a portion of the main grain-producing areas of the USA, has suffered droughts throughout history – some of the more recent in 1889, 1893–94, 1900–10, 1933–36 and 1975–76. The worst-affected states were Oklahoma, Kansas and Nebraska. That of 1933–36 generated the name 'Dust Bowl' for the region. During the 1930s, similar problems arose in Canada, in the wheatlands of Alberta, Saskatchewan and Ontario. This also generated eco-refugees.

During the pre-Dust Bowl years, the Great Plains of the American Midwest underwent agricultural expansion, then suffered drought in combination with careless farming practices, and experienced severe land degradation (an all too familiar 'boom'–drought–'bust' pattern). There is some evidence in lake cores from North Dakota that similar persistent droughts have recurred for at least 2500 years.

During the Dust Bowl years (1934–36), roughly 40 per cent of the land normally sown had to be left unharvested. Even in non-drought times it is not unusual for 10 per cent of farmland to be left unsown. Soil erosion stripped the topsoil away and led to dust storms ('black rollers'), which caused respiratory problems locally, and were still a nuisance 3000 km away. Yields of wheat fell by *c.* 29 per cent, and between 1934 and 1936 Kansas and Oklahoma lost around half of their population to out-migration (Burroughs, 1997: 45–51). A Democrat government (elected in 1932) realized that at least some of the problem that was unfolding was related to *laissez-faire* economic policies and began to intervene, which helped reduce the effects of the drought. One can argue that the livelihood and commercial exploitation strategies adopted made the region more vulnerable to its recurrent droughts and winds. The New Deal aid given by the Roosevelt government shared the impacts across the whole of the USA and allowed the Dust Bowl to recover.

Sources: Text adapted from various sources, including Worster, 1979; Bonnifield, 1979; Burroughs, 1997; map redrawn from Barrow, 1991: 174, Fig. 8.9.

Note: The 'boom' periods are when world grain prices are high; similar patterns of 'boom'–drought–'bust' were experienced in the USSR's 'Virgin Lands' of Kazakhstan (settled between 1954 and 1962), which ran into drought between 1954 and 1958, and in Western Australia's wheatlands (as well as in New South Wales, Australia).

blamed on climate change, but there has increasingly been a move to attribute problems to human misuse of resources followed by drought (the Dust Bowl 'boom–drought–bust' cycle mentioned earlier). Reliable information on the causes, severity and extent of drought and desertification is often scarce. Aid agencies are more likely to attract funding if the cause of the problem is seen to be human and 'solvable', rather than natural and something that developers will have to adapt to. Administrators prefer to blame the weather or the peasantry, rather than the development policies they have promoted. Poor nations might claim desertification to attract funding for environmental improvement and drought avoidance or mitigation projects. The latter are often simply large commercial agriculture projects that do little to alleviate the impact of drought.

The degradation of dry environments and drought-prone areas has received a great deal of attention from environmental historians, political economists, geographers and others (Blaikie and Brookfield, 1987; Barrow, 1991: 141–78; Glantz, 1994; Millington and Pye, 1994; Johnson and Lewis, 1995). There have been a number of attempts to assess the cause of past drought and famine in West Africa: Van Apeldoorn (1981) examined the situation in northern Nigeria during the Sahelian drought of the 1960s to mid-1970s; Dolby and Harrison Church (1973) reviewed drought in Africa; environment–human linkages associated with land degradation have been studied for Mediterranean countries by a number of environmental historians (for a review see Grove, 1996) and also the Thar Desert (Rajasthan, northwest India and Pakistan), (Mooley and Pant, 1981).

The reality is that large parts of the world have variable weather with periodic or occasional dry spells sometimes developing into droughts. Developers, and sometimes settlers, gain a sense of false security during wetter phases, which they tend to regard as 'normal', and view the drier periods as 'abnormal'. Ideally, activities in drought-prone areas should be conducted with an awareness of the limits faced if rains fail. Indigenous people had generally evolved adaptations to the 'boom-and-bust' productivity of harsh lands; however, modern development pressures have often broken those down, so the problem has increased.

There were severe droughts in sub-Saharan Africa between AD 1620 and 1622, causing movements of people. In all probability earlier and subsequent droughts have played a significant part in stimulating migrations in Africa and other drought-prone continents. Drought has repeatedly affected tropical, sub-tropical and temperate latitudes. Even the humid tropics have been affected, there are signs of repeated drying and associated forest fires in Amazonia, Central America, Equatorial Africa and Indonesia. Sumatra and Kalimantan had serious fires in 1877/78, 1914, 1983 and 1997. During the last few years there has been increasing evidence that ENSO events play a key role in triggering Sahelian and other droughts around the world. If knowledge about ENSO can be improved enough, monitoring the development and progress of these events may help drought early warning and identify areas at risk so that contingency planning can be undertaken.

Earthquakes

Throughout history earthquakes have taken a toll. The effects vary from locality to locality – even with the same building types (for example, solidly built Greek temples a few kilometres apart in Syracuse, Sicily, in one case have been virtually destroyed, yet another, just a few kilometres away, survived virtually intact). Earthquakes can destroy towns and city-states, obliterate libraries and infrastructure, and may cause huge fatalities. There are Biblical accounts of the destruction of Bronze Age cities, such as Sodom and Gomorra, reputedly lost without trace from the shores of the Dead Sea. These may actually have existed and were swallowed up by soil liquefaction and landslip caused by seismic disturbance about 4000 BP. Some earthquakes have caused huge numbers of fatalities; that which struck the Provinces of Shanxi, Shanxiad and Henen (China) in AD 1556 killed roughly 830,000 people, a significant percentage of the total population. Earthquakes may also initiate tsunamis and mega-tsunamis, which may destroy cultures in low-lying areas. There are also non-physical impacts, in Europe the Lisbon Earthquake of AD 1775 had considerable effect on thinkers, shaking their view that nature was benign, little-changing and generally stable.

Volcanic eruptions

There have been none of the huge fissure-type eruptions with extensive outpourings of lava and gases since humans evolved (those in Iceland in the eighteenth century may have had far-reaching effects, but are hardly large in comparison with prehistoric events). Eruptions can also project gases, dust and aerosols high into the air, even the stratosphere. Such eruptions can be violently explosive and cause blast damage, ash and debris fall-out, incandescent gas and ash clouds (pyroclastic flows or nuée ardent), acid deposition ('acid rain') and tsunami impacts. Human experience is restricted to a few, not especially large (by longer-term geological standards), explosive eruptions and more numerous minor explosive and fissure-type eruptions. The main impacts on people have been through ash-falls, pyroclastic flows and tsunamis (lava flows are usually slow enough for people to save themselves). There have probably been gas and acidic aerosol clouds at low altitude (dry fogs) without people being aware that an eruption (or marine outgassing) was the source. Some historical reports of dry fogs blame them for widespread crop and livestock deaths and human illness.

Some researchers see proof in ice cores and other lines of evidence that volcanic activity alters climate; however, sceptics question this correlation, noting that eruptions may provide a convenient, but not necessarily correct, explanation (Kondratyev, 1988: 74). Studies to check whether eruptions like that of Laki (AD 1783, Iceland) and Tambora (AD 1815, Indonesia) – examples of a fissure eruption and an explosive eruption respectively (both emitted a lot of acidic material to high altitude, and therefore are widely believed to have changed the climate) – found little proof that they did have much effect (see http://www.aber.ac.uk/jpg/volcano/lecture6.html, accessed 23 June 2000). There is a need for caution when claims are made for links between eruptions and human affairs.

History and archaeology have confirmed volcanic destruction of a number of settlements, and perhaps whole cultures. For example, the Thera eruption (possibly in 1628 BC) in the Aegean is widely believed to have initiated the fall of the Minoan civilization; Pompeii and Herculaneum were unquestionably obliterated by Vesuvius in AD 79 – the event was recorded by various chroniclers and the sites have been excavated in detail (McGuire *et al.*, 2000).

Sea-level changes

In 1931 a trawler fishing the Leman and Ower Bank *c.* 36 metres beneath the North Sea recovered a lump of peat containing a deer antler-tipped spear (of a type not used to hunt marine mammals). Since the peak of the last glacial, around 18,000 BP, sea-levels have risen as much as 130 metres.

Clearly, huge areas of land have been submerged in prehistoric times. Ports and towns founded in the Crimea between 500 and 300 BC are now under water, as is the old city of Alexandria built by Antony and Cleopatra (the cause might have been coastal erosion or the liquefaction of unconsolidated deltaic sediments during floods or earthquakes, rather than simply rising sea-level).

Even in historical times, sea-level rises have not all been smooth and gradual, there have been 'jumps'. One of these occurred *c.* 400 BC (and left a clear record in the eastern Mediterranean), and others happened around AD 1362 and AD 1634 (leaving evidence in western Europe and the UK) (Pearson, 1978: 58). The production of much of the salt used in Europe, a vital commodity, was by seawater evaporation in low-lying coastal sites. During the fourth century BC and the fourteenth, fifteenth and sixteenth centuries (AD) sea-level rises flooded salt-evaporation works in many parts of Europe, including the Norfolk Broads (UK). The resulting rise in salt prices was a serious blow for people dependent upon it for dietary needs and preserving food for winter; political unrest seems to be the result in parts of the UK, France, Ireland and Friesland (Pearson, 1978: 57).

Similar sea levels to the present day were reached during the seventeenth century AD, and there has probably been less than 30 cm eustatic variation (actual movement of the sea surface) since then in most parts of the world (some localities may have undergone subsidence or uplift as a consequence of human activity through boreholes removing water or oil, or because of mining or natural tectonic uplift relative to sea-level). Tectonic uplift has been marked in many high-latitude regions as the land rebounds after loss of thick and heavy ice cover at the end of the last glacial. So, localities may differ in the way they have been affected by the same global sea-level rises.

Diseases

History suggests that epidemics sometimes follow changes in environmental conditions or the movements of people (possibly prompted by natural causes). Contact after AD 1492 unleashed Old World diseases on the societies of the New World, against which they had little immunity. The easy conquest of the New World has puzzled historians, who just could not accept that a few European settlers in North America and a handful of Spanish conquistadors could have the vast impact they did, even with firearms and horses (Raudzens, 2001). Peoples of the New World, they argue, were decimated by smallpox, measles and other 'European' infections, together with the spread of diseases like yellow fever (probably transported to the New World from West Africa on European slave ships). Those who survived were then spiritually crushed and disorientated by the colonists; disease was effectively 'the cutting edge of imperialism' (Crosby, 1972; 1986; Diamond, 1997). The exchange has been unequal: the benefits of gold, silver and new crops like potatoes, beans, maize, tobacco, cocoa, rubber and cassava spread to the Old World with few serious diseases, while the New World had little material gain other than wheat, the horse, coffee and Asian rice. Furthermore, the acquisition of quinine from the New World helped Europeans colonize Africa, Asia and the Far East.

The case has been made that depopulation east of the Black Sea between the 1340s and 1400s (AD) as a consequence of the Black Death, cut the established source of slaves, and played a major part in triggering the trade from Africa after the 1500s. Old World diseases have helped determine the settlement and development of the New World. Even into the early twentieth century yellow fever was a major constraint on development, hindering work on the Madeira-Mamoré railway in Amazonia and causing huge loss of life during the construction of the Panama Canal. The discovery of its mode of transmission, and the development of vaccines has greatly reduced the impact of the disease (see Chapter 5).

Before the 1830s in the UK, and into the twentieth century in parts of Europe, malaria influenced settlement patterns and family fortunes. Control is generally attributed to the use of quinine; however, there may be environmental and social factors that have been overlooked: the spread of acid deposition following the Industrial Revolution and better housing (with windows to exclude insects) could have discouraged mosquito breeding and disease transmission as much as natural environmental factors.

Key reading

Fagan, B. (1999) *Floods, Famines, and Emperors: El Niño and the fate of civilizations*. Basic Books, New York (NY) (also Pimlico Press, London).

Fagan, B. (2000) *The Little Ice Age: how climate made history 1300–1850*. Basic Books, New York (NY).

Lamb, H.H. (1988) *Weather, Climate and Human Affairs*. Routledge, London.

Mannion, A.M. (1999) *Natural Environmental Change: the last three million years*. Routledge, London.

4

Are twenty-first-century humans more vulnerable to nature?

Over the long haul, *Homo sapiens* has enjoyed great biological success on the strength of adaptability. So have some species of rat.

(McNeill, 2000: xx)

Chapter summary

This chapter deals with human vulnerability to environmental factors. It is argued that modern peoples are in many respects more vulnerable than ever before; governments and planners tend to overlook the need to counter such vulnerability, above all by improving adaptability.

At the start of the twenty-first century the West is still in the throes of a paradigm shift. This began in the early 1970s when photographs of the Earth taken from space were being published in newspapers and on television prompting many to feel 'at last humans really are in control'; others began to fear that we live on a small, vulnerable planet and are not taking sufficient care of it (Ward and Dubois, 1972). While there may be growing concern about environmental problems (some say insufficient), there is too little awareness of human vulnerability to nature. Since the 1840s modern agriculture and communications appear to have saved the richer nations of the world from the famines that frequently occurred earlier, but is the threat permanently removed, and should there be more concern about other dangers? Compared with the long span of the LIA (see Chapter 3), environmental conditions have for the last century and a half been very favourable. They may not remain so.

4.1 Vulnerability

The modern world does enjoy some improvement, compared with conditions before the 1850s. In the past most people who relied on agriculture lived 'near the edge': the energy they expended to obtain food and other commodities left little surplus and, even if neighbouring countries were able and willing to offer aid, communications were often too poor to transport it. Modern agriculture and communications have helped remedy those ills. However, the risk is that people will assume that progress is sufficient and sustainable. Fagan (2000: 102) argued that

> Surprisingly few archaeologists and historians have had the chance to observe subsistence farming at first-hand, which is a pity, for they fail to appreciate just how devastating a cycle of drought or heavy rainfall, or unusual cold or warmth can be.

Modern agriculture, technology, communications and improving forecasting should have made natural disasters less of a threat; unfortunately, as well as population growth, there is also increasing environmental degradation, loss of biodiversity and risk of human-caused environmental change. The loss of life to natural disasters, and the economic losses caused by nature seem to be increasing. This is partly because the material value of properties has generally risen and at-risk populations have grown. Nature, society and technology interact to determine vulnerability. It is affected by factors like food reserves, political marginalization and misuse of power – often there is a close correlation between vulnerability and the inability to obtain basic human rights (see Box 4.1).

'Vulnerability' has been defined in various ways, as the following three examples show.

1. The characteristics of a person or group that affect their capacity to anticipate, cope with, resist and recover from the impact of a **natural hazard** (Blaikie *et al.*, 1994: 9–10).
2. The degree of exposure of groups of people or individuals to stress as a result of the impact of environmental change (Watts and Bohle, 1993).
3. The degree of susceptibility, which is partly a matter of location, partly of lifestyle (Lewis, 1999: 5).

Vulnerability is affected by factors that include: socio-economic status, ethnicity, gender, disability and age. It is an overall condition – a group of people may face more than one hazard, and the degree to which they are affected and recover is a variable. An element of vulnerability is *resistance* – the ability to withstand a hazard – and another is *resilience* – the ability to recover. 'Risk' has been defined as the likelihood of an event coinciding with elements that can be affected. Expressed as a crude equation this takes the form:

Risk = hazard × vulnerability

Box 4.1 Factors that have increased human vulnerability to the environment

- Increasing dependency – services, food production and many other things are increasingly supplied from outside a community. A problem with a supplier can have worldwide impact.
- Biodiversity damage – is reducing options for maintaining agriculture, pharmaceuticals production and so on. Human activities have been depleting natural biodiversity, and the range of traditional crop and livestock varieties used.
- Global (anthropogenic) environmental change – caused by pollution and vegetation clearance (warming, acidification, environmental pollution and stratospheric ozone loss).
- Terms of trade and agricultural development strategies that may discourage growing and storing sufficient food reserves and production of crops subject to market fluctuations.
- Communication, settlement, social and healthcare developments – increase risk of disease – rapid travel, crowded cities, misuse of antibiotics, and the concentration of migrants and refugees who are weakened and vulnerable.
- Weapons of mass destruction for armed forces and terrorism – potential to dislocate and debilitate populations and disrupt the environment.
- Increasing human population – stress upon resources and environment.
- Intensification of agriculture, industrial and medical use of biotechnology – risk of new 'out of control' challenges.
- Populations who have little 'survival talent', only specialized skills – these people would find adaptation difficult.
- Complex technology and livelihood strategies – these are easy to disrupt and difficult to service, repair or replace.

Many of today's humans have become less adaptable than ever before: people in richer countries are less able than were their ancestors to cope with inadequate food, contaminated water and harsh climate, and are vulnerable if modern services and infrastructure become disrupted. There is also less chance for an individual or family to be independent when most developed-country livelihoods are complex and depend on many others around the world. People in developed countries tend to have few children and smaller, less extended families than in the past, so in time of crisis there is less chance of help from a relative (Couper-Johnston, 2000: 75–6). In poorer countries, loss of access to common sources, population growth, increasing poverty, globalization and environmental degradation have increased the vulnerability of many people.

There are three aspects of the environment that have a particular bearing on human fortunes:

1. limits (the availability of inputs; ability to cope with outputs (e.g. pollution) and space for development)
2. potential (opportunities)
3. hazards.

Since the early 1970s awareness has grown that there are environmental limits, and that they may be getting very close or may already have been violated. The potential for sustaining development is getting no easier because demands increase as populations grow and people increasingly seek higher standards of living. Agricultural and industrial activity, warfare and unrest are stressing the environment and the social institutions that play a part in environmental management, making people more vulnerable to hazards, and even increasing the likelihood that they will happen.

Vulnerability to natural disasters has been researched by a wide range of social scientists, notably cultural geographers, and by those interested in disaster preparedness and relief. Development should be a means of reducing vulnerability, yet it often has the opposite effect. Hazard exposure is growing: people increasingly settle and establish livelihoods in hazardous locations; 'development' pressures may make them less adaptive and less resilient; and global environmental change is complicating or adding to natural threats (Dow, 1992). For example, destruction of mangrove forests increases a coastal region's exposure to storms and tsunamis, and deprives local people of a range of (free) foods and fuelwood. The shift from mixed subsistence agriculture to less diversified cash-cropping is likely to mean more vulnerability through market fluctuation, depending on external inputs, and pests and diseases risks associated with mono-cropping (Wijkman and Timberlake, 1984).

Disease epidemics are the result both of natural factors and human actions. Watts (1997) argued that European imperialist policies are to blame for many of the epidemics of the past 500 years. Such (simplistic) social determinism discourages monitoring of environmental factors; although, it also warns of the need to assess carefully what impact the complex process of development has upon ongoing natural threats. Many things can damage or destroy established 'safety nets' and either tempt or drive people into risk situations; so reduction of vulnerability demands a multidisciplinary and multisectoral approach. Lewis (1999: 40) argued that measures to reduce vulnerability are more cost-effective than measures to try and protect against a hazard.

According to Wijkman and Timberlake (1984: 23), there were around 22,570 deaths worldwide attributable to natural disasters in the 1960s; this rose to 142,920 during the 1970s. In the 1970s roughly six times more people died as a result of natural disasters than in the 1960s, but the number of disasters had only gone up by about 50 per cent; these trends have probably continued (Wijkman and Timberlake, 1984). A recent assessment of disasters and the international response is presented by the International Federation of Red Cross and Red Crescent Societies (2001), which suggested that roughly 300 million people were affected by natural disasters in 1991. This report stressed that people are nowadays often more vulnerable to natural disasters than they were in the past (although from the mid-nineteenth century to the present less vulnerable to famine in developed countries).

While human vulnerability to environmental threats has grown, this is offset a little by the availability of better forecasting, modelling, hazard and risk assessment, and improving communications. There are also likely to be developments in fields like extraterrestrial body strike monitoring and mitigation, and seismic and volcanic threat forecasting. Humans, especially since the 1830s, have altered the Earth's environment more than any other organism since cyanobacter and algae changed the primordial atmosphere roughly 2500 million years ago (McNeill, 2000: 265). Humans are causing one of the greatest 'Gaian' changes, and are doing so with little thought, without knowing all the 'crucial rules' about biosphere maintenance, and ignoring much of the little they are aware of; all the while doing their best to 'always look on the bright side'.

Vulnerability assessment has grown up to assist various fields of planning, it can draw upon tools like GIS, hazard mapping, social survey and risk assessment (Lewis, 1999). There are flood vulnerability maps for many of the world's river systems, and some governments now run national or participate in multinational flood early warning and mitigation schemes. Some mountainous countries have developed avalanche and landslide vulnerability assessment; where bushfires are a problem hazard mapping is increasingly used; Japan and California use earthquake hazard assessment. These assessments have generally neglected the social aspects of vulnerability. Yet, in many situations, 'special needs' groups are increasingly at risk, and find escape and recovery from natural disasters less easy (Cutter, 1996; Cutter *et al.*, 2000). An example is Tokyo, which has a considerable elderly population who are more likely to live in vulnerable wooden houses and older high-rise apartments, and consequently suffer more in earthquakes (Takahashi, 1998). Unfortunately, many countries only use hazard assessment when conducting impact assessments for large construction projects, which gives very patchy coverage.

The age of buildings, their height, mode of construction and location has considerable impact on their vulnerability to earthquakes, storms and floods. Mexico City has experienced a number of building phases: pre-1960s sturdy (and lower-rise) construction resists earthquakes quite well; the 1960s and 1970s expansion boom generally meant weaker construction of intermediate height, which suffers more from seismic damage; post-1970s buildings have been subject to tighter building regulations and withstand shocks better than those of the previous few decades. Where buildings are located also matters: for example, some Mexico City buildings are secure on solid ground, others have their foundations in poorly consolidated former lake deposits and these easily liquefy during earthquakes, ensuring that almost any type of construction sinks or topples. In Fiji, modern lightweight housing is often less resilient to storms than traditional housing, and is much more costly to repair.

While poverty does not automatically mean greater vulnerability it is often the cause. The rich can often afford to live in the cleanest, healthiest and safest places – *if* they realize the risks. They are also more likely to get warnings of storms and other threats, and have the necessary vehicles to make an evacuation. The poor are likely to be in flimsy housing sited where there is flood risk, landslide risk or vulnerability to storms; their lack of education, inability to afford insurance, unwillingness to flee insecure homes that may be looted, and limited access to transport means they are seldom effectively evacuated. Poor housing is more likely to collapse, causing injury and economic loss. A frost or snowstorm that has little impact on rich urban dwellers may strike the poor who live on the streets or in shacks. There may be exceptions, however, as the following examples show.

- Blaikie *et al.* (1994: 61) note that Turkish 'guestworkers' in Germany tended to return home and spend their savings on bigger houses made of traditional materials, which were often more vulnerable to earthquakes than those of their poorer neighbours.
- Rich people sometimes buy homes in vulnerable places to enjoy views or beach facilities.
- Poor people, compared with the rich, may be more streetwise and resilient in the face of disaster.

The poor of urban areas are vulnerable to the diseases of crowding, poor infrastructure and services: influenza, TB, dengue, rodent-related infections, and the effects of air and sewage contamination. Cities are much less epidemiologically isolated nowadays than ever before: in the late 1940s about the fastest trip a dozen or so (rich) travellers could make from Cape Town (South Africa) to Southampton (UK) would have been roughly eight days by unpressurized Imperial Airways flying boat; today the flight time is around 12 hours (the cost far more affordable), and over 300 people do it confined in an aircraft that recirculates the air they breathe. So diseases can now spread more easily – in the past there was just about time for them to manifest and give the authorities a chance to contain transmission. Even in richer nations there has been environmental degradation and practices that help diseases 'jump' between people and from other organisms to humans; these include drug abuse and needle sharing, and amongst the 'underclasses' growing poverty and overcrowding.

Trade carries all manner of goods, and often stowaway organisms, around the world to an extent never seen before. Cholera introduced into Peru spread through Latin America in 1991 after a ship from Asia pumped contaminated ballast water into Callao harbour: over 300,000 people soon caught the disease. Over 500 million people cross international borders each year by aircraft. Often, these travellers have been visiting countries where inadequate water supplies, sewage disposal and healthcare are getting worse year by year, and where civil unrest and refugees are all too common. Add to this the problems of poor insect control and misuse of antibiotics, and there is a growing chance of disease spreading, and the transfer of pathogens that have been gene-swapping and developing drug and antibiotic resistance. An American microbiologist has described the twenty-first-century world as becoming a 'viral superhighway'. Diseases have always evolved, but today humans are aiding disease mutation and making transmission easier.

Aarmelagos (1998) suggested there have been three major epidemiological transitions.

1. Around 10,000 BP, when people shifted from nomadism to sedentary farming. This meant there was denser population, proximity to livestock and farm pests, and sustained contact with waste and infected sites. All this led to new human diseases or greater prevalence of some existing diseases like scrub-typhus.
2. Roughly 1971, when great progress had been made against smallpox, polio, TB and malaria.
3. In spite of greater life expectancy at birth in developed countries than ever before, vulnerability has probably risen post-1980, since when there has been the appearance of 'new' diseases (HIV/AIDS, BSE/CJD), the return of 'virtually conquered diseases' in a more virulent form (drug-resistant TB, malaria and venereal diseases) and an increase in 'diseases of affluence' (notably coronary problems, cancer, asthma and pollution-related allergies).

Aarmelagos (1998) stressed that, while civilization today does expose people to diseases, it also offers improving medicine, better monitoring, education and sometimes good governance. Much of the threat we face could be reduced *if* governments were vigilant, invested in monitoring and preventative measures, and maintained and improved healthcare and public services. Risks can be reduced if funds are made available and if management is good – failure to do these things could mean catastrophic loss of life. History has shown just how bad serious epidemics can be, notably Black Death, smallpox and cholera outbreaks in Europe between the Middle Ages and the twentieth century.

Urban development is exposing both rich and poor people to risk, but especially the latter. Uitto (1998) argued that urbanization, especially the growth of mega-cities was among the major global challenges faced today. In recent years world demography has witnessed a shift to the point where more than half the world's people are living in urban areas and the proportion is growing. These people are increasingly out of touch with nature, and living in largely artificial and often unhealthy environments. People in these urban environments are *not* as safe as they might like to imagine from adverse environmental conditions. Large cities are complex ecosystems, with much that can be disrupted, and once disrupted their recovery can be a struggle during which large numbers of people suffer.

Unfamiliar with nature and the ways in which a city obtains its water, fuel and food supplies, urban people may lobby and vote for policies that increase their vulnerability, and possibly also make life difficult for rural people. They are unlikely to give much support to rural development issues unless these benefit urban food and water supplies or environment. Large cities can also be like astronomical black holes, increasingly attracting more than their fair share of national investment and provision of services. In many developing countries, which used to be largely self-sufficient in food production, cheap grain is imported partly to keep down prices for urban consumers, and this has had the effect of helping to ruin farmers in the countryside. This in turn leads to livelihood breakdown and environmental degradation as non-sustainable herding or marginal land clearance takes over; parallel with this is reduction of indigenous food production and dependency on overseas suppliers.

There are now more rapidly growing mega-cities: in 1900, there were 11 metropolitan areas with populations of over a million (mostly in what are now the developed countries), today, there are over 400 cities of more than a million people; of these 28 are mega-cities with a population of over 8 million and some are huge. In 2000 Mexico City probably had a population of 20.1 million, Tokyo 28.0 million and São Paulo 22.6 million (Fuchs *et al.*, 1994). Even those mega-cities that are not growing especially rapidly are facing challenges in obtaining water, energy, disposal of waste, and the management of traffic and air pollution. Keeping up with the growth to ensure an adequate infrastructure is a problem, and often the inhabitants are poor, unemployed or under-employed (Mitchell, 1995; 1999; Parker and Mitchell, 1995; Mitchell, 2001; Hardoy *et al.*, 2002). Growing urban populations mean it is increasingly a challenge to control people, feed them and prevent disease outbreaks in the event of a natural disaster or other emergency.

Presently, roughly half of the population of the world's 50 biggest cities are living within 200 km of geological faults that are known to suffer earthquakes of Richter 7 or greater. Uitto (1998) calculated that this is about twice the level of exposure there was in 1950. Some of the mega-cities at risk have a far-reaching influence on the world economy. The obvious examples are Tokyo and San Francisco; a major earthquake hitting one of these would be felt in stockmarkets and manufacturing around the world (Hadfield, 1991; Shah, 1995). The recent Kyoto earthquake reduced the production of integrated circuits ('chips'), causing problems for the electronics industry worldwide, and a market shortage is still encouraging the theft of computers or their component parts in various countries. So, globalization of trade and manufacturing is tending to increase vulnerability worldwide.

4.2 Living closer to the 'limits'

The twentieth century has been marked by several major wars, numerous smaller conflicts and, in some regions, widespread lawlessness. In Africa, in particular, since the 1960s some 30 million people have been forced to relocate by warfare and unrest. Displaced people, if they are not dispersed, are likely to destroy resources and wildlife, and may spread disease to other populations; they are also poor and vulnerable to natural disasters.

Unrest wastes resources, and generally diverts skills and funds from more useful activities. In addition to actual conflict for most of the second half of the last century there was a Cold War that squandered resources on superpower rivalry, and hindered the study of the Earth's structure and function. In the twenty-first century the capacity to cause widespread nuclear destruction is slightly diminished but genetic manipulation has added to the arsenal of chemical, biological, nuclear and conventional weapons, and terrorism has become a greater threat (since writing this in draft form the New York World Trade Center tragedy has underlined the risk).

As well as an ongoing threat from natural environmental change and sudden disasters, many environmentalists claim there is a global development-related environmental 'crisis', and that it is affecting the poor of the developing countries disproportionately more. A crisis is a turning point, the last chance to avoid or reduce damage, or to salvage as much as possible. Sometimes a crisis can offer opportunities. The problem is to prove whether or not there is a crisis and, if there is, to establish how long people have got to 'get their act together' and what they should do. Environmental impact assessment (EIA) has become a widely used tool since 1970; *vulnerability impact assessment has not been given enough attention.*

There have been warnings that the technology being used to grow enough food to feed the world's expanding population is leading to soil degradation, an increasing shortage of clean water, and environmental pollution. Technical innovations have altered food production to more than keep pace with population increase, so far; but such improvements are likely to be 'masking' ongoing and serious damage, and the point will come – if nothing is done to reverse this trend – when breakthroughs will no longer appear or keep up with demographic increase. Worse, what production we enjoy at the moment might decline through pollution, soil degradation, global environmental change, pollution, new natural challenges, and social or economic upheavals.

Some people fear that 'high-tech' food production is vulnerable to pests, pollution, diseases and climate change. If these fears are correct agriculture methods will have to change. Food production has increased *c.* 24 per cent per capita and fallen in price *c.* 40 per cent since 1961, but is it sustainable? In 1972 failed harvests forced the USSR to buy large amounts of grain from Canada and the USA; intelligence agencies subsequently joined the large food and commodity trading companies and the UN Food and Agriculture Organization (FAO), to monitor current and forecast future production and demand (CIA, 1974). In the UK and Europe, some might say there have already been signs of 'too high a price for cheap food' and of excessive vulnerability of production. The manifestations are: the spread of BSE; the

2001 foot and mouth disease outbreak; environmental contamination associated with aquaculture of salmon and other fish; and the need to set aside farmland to try and reduce nitrate pollution of groundwater.

Land degradation and drought impact seems to have increased in West Africa, North Africa, Malagasy, Ethiopia and numerous other countries. The removal of vegetation cover through overgrazing, exploitative farming and fuelwood collection have played a part in this. As population grows, agriculture might place further pressure on the environment until a threshold is passed and crop yields fall markedly (Voortman, 1998). When sedentary agriculture began to develop, roughly 8000 years BP, the total world human population was somewhere between 2 and 20 million; by the first century AD it had risen to 200 to 300 million. Today the latter figure would be roughly equivalent to the population of Indonesia alone (McNeill, 2000: 8). So there are serious challenges to be faced by food production over the coming few decades, as outlined below.

- Human population, currently around 6.5 billion, may well rise to 8 billion by AD 2020.
- The intensification strategies used since the 1950s to boost agricultural production have come near to their limits – spectacular yield increases seen in the 1960s and 1970s are much rarer nowadays, and the agrochemicals and mechanization involved are causing environmental and socio-economic problems.
- There are growing difficulties caused by acid deposition, reduction of stratospheric ozone, tropospheric ozone pollution and other air pollutants.
- Increasingly there is competition for and contamination of water supplies.
- Soil degradation (erosion, structural damage, contamination and fertility decline) has outstripped soil formation in most farmed areas.
- There are likely to be problems caused by global warming.

The world is increasingly dependent on artificial (chemical) fertilizers. Much of what is applied ends up polluting the environment, rather than feeding crops. Some countries, like The Netherlands, the UK and parts of the USA, have had to legislate to reduce chemical fertilizer usage. The use of these fertilizers and modern seeds has helped to 'hide' soil degradation by keeping yields from declining, even though the land is suffering structural and chemical damage. Recent estimates suggest that humans have increased soil erosion rates by two to three times natural rates; farmland is also being lost beneath urban sprawl and other infrastructure. In the early twenty-first century, over one-third of the Earth's surface is suffering significant land degradation (McNeill, 2000: 58–9). The diversity of food sources has been falling, nowadays probably around two-thirds of the world's food is obtained from three grain crops (rice, wheat and maize). Worse, the number of varieties of these cereals has been cut back by commercial forces and agricultural development policies. Humans are very dependent on a limited range of grass species (a worldwide situation uncomfortably like that of Ireland before the Potato Famine started in 1845).

The challenge is to increase and diversify agricultural production, and to sustain it, without much more new land and in the face of environmental challenges, and to do so without the levels of pollution and soil degradation caused so far. And, do all this in a couple of decades in a world where warfare, unrest and economic disruption are ever-present threats, not to mention greenhouse warming and natural disasters. There have been calls for a '*Doubly* Green Revolution' to meet these challenges, something twice as effective as the **Green Revolution** of the 1960s and 1970s, but much less environmentally damaging. While pointing out that there can be no single recipe for such a revolution, Conway (2000: 15) argued that it could be done, and outlined what he felt were the essential ingredients. One of the key elements is likely to be a cautious 'poker game' with GM techniques (genetic engineering and biotechnology) as the high-value cards, and environmental disaster and hunger the price of misjudgement or non-action.

4.3 Critical regions?

As already discussed, there are social groups today that are more vulnerable than others. It is also possible to look at vulnerability spatially and to recognize regions and resources that are at particular risk (Kasperson *et al.*, 1996; Kasperson and Kasperson, 2001). Some of these regions – such as the wheatlands of North America, which produce much of the world's grain surpluses – may be especially critical to human wellbeing worldwide. In India the advances achieved through the Green Revolution since the 1960s have so far kept famine at bay, even though there have been droughts and other environmental disasters that in the past would have caused hunger. However, India's Green Revolution has led to two states dominating the production of grain surpluses: Haryana and the Punjab. In both of these areas there are now signs that the environment is suffering from agrochemical pollution, soil degradation is getting worse and the situation could be said to be swinging toward critical; without major changes to agriculture soon, future environmental challenges may not be met with adequate food reserves.

The criticality of urban, especially some mega-city, regions has already been outlined; other vulnerable areas are the various 'marginal lands', where current land use and little or no environmental challenge will result in a threshold being passed, beyond which there is likely to be severe breakdown of agriculture, water supply, regional economy and social cohesion. The semi-arid lands of the developing countries stand out as such areas. In a number of countries rural people unable to cope with change have either become rural–urban migrants or have opted to resettle in less attractive marginal lands (often drylands) (Glantz, 1994: 20). The commercialization of agriculture since the 1960s has meant that better land has often been acquired by large commercial or private landowners, forcing smallholders to move to more marginal environments, a familiar pattern in areas like northeast Brazil, Sahelian Africa and many other regions. Consequently some marginal lands have had considerable population increase through in-migration, and probably also as a consequence of medical innovations like smallpox control.

Pretty and Ward (2001) argue that there has been too little study of how social and **human capital** affects environmental outcomes. Social capital may be defined as the value added to any activity by human relationships and cooperation, although agreeing a definition acceptable to all is a chal-lenge (Paldam, 2000). Social capital is vested in social institutions (families, extended families, communities, guilds, cooperatives, associations, busi-nesses, NGOs, schools and trades unions), relations of trust, reciprocity, common rules, norms and sanctions. Not all forms of social capital are good for everyone – some may be based on fear, rather than trust, such as feudalism, mafia activities or slavery. Also, while social capital may be 'the glue that makes people work together', 'glue' can be the reverse of a lubri-cant when it comes to social change and innovation. This was the case in the UK in the early 1970s, when trades unions hindered adaption to changing world conditions (Paldam, 2000: 635).

Changes in beneficial social capital, especially its breakdown, help deter-mine the way in which the environment impinges upon humans, and how humans affect the environment. There are numerous documented cases where breakdown of local institutions has led to environmental degrada-tion, especially where common property resources are being used. As local institutions have withered, the state frequently takes over and, all too often, is out of touch and unable to manage as well as the locals once did. Pretty and Ward (2001: 211) pointed out that there is a growing need for forms of social organization suitable for natural resource management and protec-tion at the local level – but it is seldom possible to simply revive old institutions and traditions once these have been lost.

Some countries have high rates of HIV/AIDS, and the signs are that the problem is on the increase. The consequences are growing numbers of single-parent families and orphans, often also infected. These pose a burden on healthcare, give rise to increased welfare demands and social disruption, and in some regions there is already a shortage of labour to sustain agricul-ture.

4.4 Adaptability and resilience

Adaptability and resilience have been key elements in human survival, and large numbers of people have lost or are losing these qualities (see Box

4.1). Yet, modern humans have access to technology and knowledge undreamed of even 50 years ago. Since the late 1950s there has been increasing international research cooperation and sharing of data (expressed in metric units, which can be used by almost everybody). The end of the Cold War has been marked by greater exchange of information and, hopefully, has freed up more resources for development and environmental management. Advances in computing and the development of the Internet have also aided the monitoring of environment and development issues and has made responses easier. There are some grounds for optimism. New capabilities should be focused on improving adaptability and resilience.

How can people withstand environmental and socio-economic challenges without suffering a sustainable livelihood breakdown leading to chaos and environmental degradation (there may be a change in activities – a state change – as part of the process of resilience)? *Social resilience* has been defined as the ability of groups or communities to cope with external stresses and disturbances as a result of social, political and *environmental* change (Adger, 2000). Social resilience seems to be especially important for groups and communities dependent on environmental resources for their livelihoods. If sustainable development is to be achieved and maintained, various challenges must be weathered; to do so it will be crucial that the institutional structures (e.g. property rights) are effective and flexible.

A number of researchers have stressed the need for human development to better embrace the goals of resilience and adaptability (Holling, 1978). However, much of what is done by modern decision-makers is inflexible 'juggernaut' development that cannot easily change to cope with new challenges (Adams, 1992).

Countries – especially the rich nations – are increasingly dependent on electronics. Unlike a card index with little to break down or rapidly become obsolescent, a computer system is vulnerable: hackers may corrupt software and misuse data; criminals and terrorists may infect it with viruses; there may be hidden programming errors that only become apparent during use; power surges caused by faulty generation, solar storms, nuclear weapons or cosmic radiation can cause havoc; and there is a threat from civil unrest and terrorism, floods, storms and fires (Crane, 1990). Also, computers cost a lot more to install, service and replace than card indexes. Already solar storms have wrecked national power distribution grids for weeks at a time in Canada and the USA, and periodically threaten short-wave radio communications and the satellites used for mobile and international telephone and data links, strategic early warning, navigation and earth monitoring. The largest solar flare would have minimal effect on human affairs before the 1960s but, nowadays, could cause chaos and perhaps even prompt a nuclear weapons exchange.

The degradation of disaster avoidance and coping strategies

Mobility was a common coping strategy in the past; people often simply moved somewhere better if drought or some other problem occurred. Today borders are difficult to cross, land is fenced, owned and policed, and unoccupied 'common' land that might offer livelihood opportunities is increasingly scarce. Some of these difficulties arise from modern governance, some from population growth, and some because of the expropriation of common land and common resources by corporations, states and rich individuals. The loss of common resources has deprived many poor people of important parts of their normal livelihoods and disaster avoidance strategies. Social change can also damage coping strategies; for example, traditional checks and safeguards that prevent overexploitation and ensure sustainable livelihoods may be based on taboos and superstitions. If the latter are rejected the moderating influence is lost and may be difficult to replace. The continued penetration of capitalism means that people tend to be working more and more on producing a cash crop, rather than practising a diversity of activities to survive. If there is a fall in demand for the cash crop or the production fails people will probably not have the time or skills to pursue alternatives.

Migrants, refugees and eco-refugees

Migrants may move away from their place of origin for extended periods or just seasonally, usually the move is temporary and they have some degree of choice in the decision. Refugees are dislocated unwillingly and for an indefinite period. Migrants generally maintain 'roots' and may return to family and possessions, refugees usually leave in a hurry, sever contacts and lose possessions, and are less likely to return home.

Relocated people are vulnerable because:

- they are removed from their familiar resource base (and may have little relevant local knowledge)
- they have reduced opportunities for self-reliance
- they often congregate in a hazardous location
- through trauma, stress and poor housing they are more subject to illness and disease.

The world has yet to see huge numbers of people seeking resettlement as a consequence of rising sea levels or climate change (eco-refugees), but there are predictions that this is likely within the century. Already, in many countries there has been a growing movement of people from rural to urban areas. Some of these have made the move in response to environmental problems triggered or exacerbated by various development activities, others because they seek their fortunes in urban areas for a range of reasons. The flow of such relocatees places stress on urban areas in developing countries, and increasingly on developed countries when the migrants seek asylum. Rural–urban migration is often a vicious circle, causing labour shortages in the countryside, which in turn lead to environmental problems, more relocations and further breakdown of traditions.

4.5 Surprise events

Are environmental surprises more likely nowadays? Will these surprises cause greater problems? The answer to the first question is that, in addition to natural events, environmental change caused by humans may throw up surprises. In answer to the second question, there should be better means for predicting or spotting the development of surprises; however, if the effort is not made to look, or the search is ineffective, problems will appear unexpectedly; also, with greater populations than ever before and people who may be more vulnerable, impacts may well be more likely and have greater impact.

Environmental surprise has attracted increasing interest in recent years (Kates and Clark, 1996). Streets and Glantz (2000) explored the concept of climate surprise and concluded that such events cause marked damage (perhaps the most impact) to human wellbeing, and that some kinds of surprise are truly unpredictable but others can be anticipated. People living in a particular region are often aware in a general way of the likely hazards, although they often do not know when they will occur or what their intensities will be. Such knowledge and experience can be upset when environmental change or human development shifts the zones in which hazards occur. Some surprises are things that were never thought possible, like the sudden confirmation of stratospheric ozone thinning in 1987 and the recent realization that huge tsunamis can be triggered by eruptions or earthquakes.

Many surprises can be forecast *if there is a willingness to do so*, but making a forecast does not guarantee avoidance or adequate contingency planning for mitigation; communication between expert advisers and decision-makers can be problematic. Streets and Glantz (2000: 101) gave an example of this: the ongoing risk of drought in the US Great Plains is recognized by historians and climatologists, and by staff of soil conservation services originally set up in the 1930s in response to such problems (see Box 4.2 and Fig. 4.1 for another example of a situation where authorities chose to ignore advance warnings of human-induced environmental disaster). Nevertheless, many officials and farmers, because they had less than a few decades of personal experience and were subject to business pressures, were surprised in the early 1980s by drought and land degradation, and some

Box 4.2 Creeping environmental change – the Aral Sea disaster

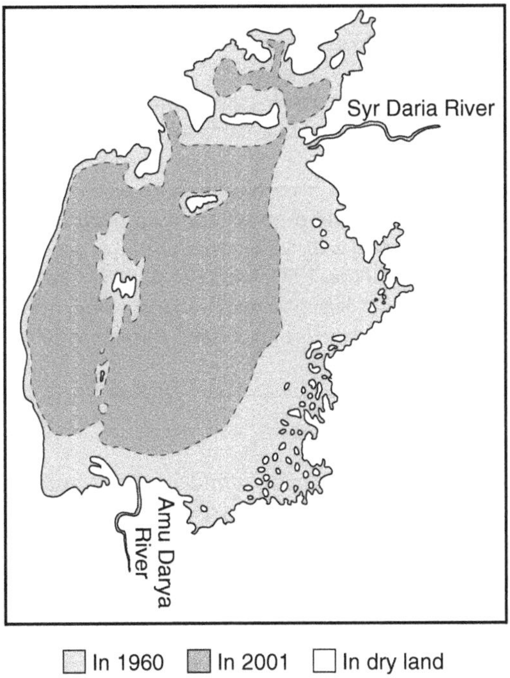

☐ In 1960   ▦ In 2001   ☐ In dry land

Figure 4.1 The extent of the Aral Sea disaster

The Aral Sea had shrunk by about 60 per cent of its area and 80 per cent of its volume by 1998 (a c. 18 m drop in sea level). In Fig. 4.1 the extent in 1960 is shown by light and dark grey; in 2001 it had shrunk to the dark grey area. There have been phases of natural shrinkage in the past; however, the main cause of water loss over the past few decades has been human activity, notably diversion of inflows for irrigation (this is then lost by evaporation). Return flows from irrigation are salty and contaminated with agrochemicals; added to this is industrial and nuclear pollution, including PCBs and heavy metals.

The loss of fisheries and wildlife has been marked – of 24 commercial fish species caught 30 years ago, only four now survive and there is virtually nothing worth harvesting. Local populations have suffered economic and health impacts – for example, cancer of the oesophagus is especially widespread, also liver disease and gynaecological problems (Mickin, 2001). Exposed areas of salts and pollutants are blowing across to contaminate surrounding regions.

Few would argue that the Aral Sea (CIS) is not a huge environmental disaster. Although it is the result of human activity the case is very relevant to natural disaster warning and avoidance. Streets and Glantz (2000) suggested that the disaster was the result of 'gradual change and the failure of the authorities to alter their perceptions fast enough'. Members of the Soviet scientific community had long recognized the

---

Box 4.2 Creeping environmental change – the Aral Sea disaster – contd

decline and had issued their leaders warnings for decades (the first in the late 1920s). It was obvious that the irrigation projects in central Asia were diverting too much water from the Amudarya and Syrdarya Rivers.

Warnings were ignored until a number of obvious thresholds had been passed. The seriousness of the decline then came as a 'surprise' to the decision-makers, yet it had 'crept up' over more than 40 years.

The Aral Sea disaster serves as a warning: in spite of clear experts' advice about threats and natural thresholds, the problem crept up because decision-makers and local people failed to perceive the threat or were unable to act.

Sources: Text sourced from Conte, 1995; Glantz *et al.*, 1993; Glantz, 1999; Goudie, 2002: 50–5; http://www.dfd.dlr.de/app/land/aralsea/chronology.html; http://ntserver.cis.lead.org/aral/level.htm (accessed December 2001); map drawn from several sources.

---

land users had even been removing windbreaks planted after the 1930s droughts. Another example is provided by citrus growers in Florida: 'old timers' were fully aware that occasional frosts can cause havoc, unfortunately many newcomers chose to plant high-yield or especially marketable varieties that were not frost-hardy, and in the 1980s a number of cold spells decimated their plantings. Not only ignorance is at play here – a cynic once observed that 'acceptable risk is directly related to the amount of profit that can be made'. People also accept higher risks if they feel they are 'in control' of the situation. Sometimes people fail to judge a threat as serious – 'it won't happen in our lifetime' – or adopt a fatalistic approach; when such attitudes are coupled with a risky situation that offers opportunities disaster may follow. Sometimes people are forced to take risks: in Bangladesh the poor have little alternative to settling the, often quite fertile, land that is at risk from storms, floods and tsunamis, because other areas are occupied.

When assessing the overall effect of a surprising event it is necessary to consider the impacts on all stakeholders over long periods. Where people are aware of a hazard they may decide to try and avoid it or prepare to reduce its impact. The trouble with adopting an anticipatory approach is that it usually costs money to be prepared, preparations may be seen to 'hinder development' and there is a risk that the forecast will be wrong. Administrators and decision-makers treat forecasts with caution because they are afraid they will be found to have 'cried wolf' and be discredited. As US President Jack Kennedy is reputed to have said, 'Success has many fathers, but failure is an orphan.' Fear of a threat may cause negative impacts – depressed land values, runs on currency, out-migration or riots – so warnings may be suppressed. An example of a warning that was widely heeded was that circulated before 2000 about the Y2K 'millennium bug'. This threat largely failed to materialize, and surprisingly little criticism has been levelled at those who forecast disaster at the hands of this programming blunder; perhaps those who paid for it are convinced that the costly software they installed worked, or are too embarrassed to admit it was largely a false alarm.

The ideal solution to issues that demand a proactive approach is to develop a win-win solution, one that pays off no matter what the outcome. So a carbon sink forest may not be needed if global warming fails to materialize, but it can still provide timber, recreation and conservation benefits paid for by a debt-for-nature swap.

At present a great deal of attention is being directed at the threat of global warming through carbon dioxide (and other greenhouse gas) emissions. The focus is on how to try and prevent the warming primarily by controlling emissions; yet many argue that the change is now under way and will inevitably get worse. If the latter view is correct, attention should shift to predicting detailed scenarios, and to mitigation and adaptation measures, rather than emissions control alone. There is also a risk that a well-publicized threat like global warming takes attention and funding away from dealing with other perhaps equally important (or greater) threats, such as soil erosion, environmental pollution with pesticides or sudden natural climate shifts.

Streets and Glantz (2000) suggested that surprising events can be divided into 'rare events with serious consequences' (e.g. mega-volcanic eruption, asteroid strike, shift of the North Atlantic thermo-haline circulation resulting in southward depression of the Gulf Stream, catastrophic surge of the west Antarctic ice sheet) and 'common events that elude detection and prevention' (e.g. severe storms, floods, unseasonable frosts). Not all surprises happen suddenly – some 'creep up'. Gradual change with no clearly identifiable danger threshold can mean that no action takes place until it is too late. For example, several wet years without bushfires can result in the accumulation of enough plant matter for there finally to be a disastrous fire. Similarly, gradual soil loss of a few millimetres a year will probably go unnoticed by locals but, given long enough, a point is suddenly reached when agriculture is no longer viable. These critical thresholds will be missed if there is not effective environmental management. A more colourful description of the more drastic consequences of passing a threshold is the physical, chemical or biological (or a combination of these) 'timebomb' – a slow build toward a threshold with a catastrophic change suddenly triggered when that threshold is reached or some factor sets things off unexpectedly. An example of a chemical timebomb is the gradual accumulation of a pollutant, say an agricultural pesticide, in soil. Bound to clay particles, the poison is held like the electrical charge in a battery; at some threshold point either so much has accumulated that the soil can no longer hold it or some environmental change (like acidification or climatic warming) leads to sudden catastrophic release, and contamination of streams and groundwater (Myers, 1995).

There had there been decades of monitoring carbon dioxide levels (for purely academic reasons) before links between past climate change and atmospheric carbon dioxide and methane were established. Now, in an era where research is increasingly funded only if there are likely commercial or social benefits, it is doubtful that such 'non-applied' studies would be undertaken (and no early warning would be gained). Environmental managers are increasingly being called upon to identify potential threats and give early warning of threshold points – often before they have adequate knowledge or data. A catch-22 situation: if proactive measures are not taken, problems become costly or even unsolvable; but researchers and administrators want to wait until threats definitely develop (and that is too late). Forecasting problems is difficult because so many, often unrelated, factors are involved. They may come together to cause a difficult-to-predict synergism (a 'cumulative impact', in the parlance of impact assessment). For example, the spread of a 'low-profile' alien species like Chinese mitten crabs to the UK and Europe may increase vulnerability to storm and rising sea-levels, because the animals riddle sea defences and river banks with burrows. Because these cumulative impacts are indirect and often complex, their prediction is a tremendous challenge – there will be future surprises missed by hazard and risk assessors.

Decision-makers, especially in democracies, tend to focus on short-term problems. Their voters are more concerned with immediate issues, and those they elect are more likely to tackle the things they feel there is a chance of resolving in their term of office. Long-term developments are seen as 'over the horizon' and someone else's problem (NIMTO: 'not in my term of office'). Citizens are more inclined to support spending on something that affects them directly in the near future and, to a lesser extent, their offspring – but distant generations and the children of those in other countries generally generate much less altruism. Since the 1940s there has been a growing attitude in the West to argue for no action against a threat that might strike in 10 or 20 years and to wait in the hope that 'technological advance will come to the rescue'. Unfortunately, technological advance takes time and is dependent on foresighted decisions being taken; when under way progress is usually in a series of steps and plateaux – the steps may be too few and too shallow to counter a problem without other developments (not necessarily technological). So far, food production improvements have largely kept pace with rising population, however, one cannot assume a steady progression of breakthroughs; worse, climate change, social unrest, accumulating pollution and other factors may conspire to cause stagnation or even decline of production.

People tend to cling to their beliefs and expectations, and hope that conditions they have become used to will continue. Faced with something like climate change or drought, those with such a mindset see it as 'abnormal' and may delay adaptation in the expectation that things will return to 'normal' (this may have been a factor in the demise of the Norse settlements in Greenland).

Minimizing surprises depends on a complex mix of better research, improved planning and decision-making, and governance that enables spending on forecasting and advance action. Research funding tends to flow along well-established channels; Streets and Glantz (2000: 107) argued that there should be a move from developing more knowledge about what we already know, toward learning more about 'black holes' – gaps in our awareness.

4.6 Climate change vulnerability

During the 1960s and 1970s a number of climatologists warned that there might be a risk of global cooling. Nowadays, few fear the onset of a new 'ice age', and many worry that global warming may be under way and accelerating. Are we worrying too much about what we are doing to the environment and not enough about what the environment can do to us? *It is better to focus on improving the ability of people to adapt and on vulnerability reduction than on mitigation, because the former strategies give some pay-off whether disasters are natural or caused by humans, and should have some effect against unexpected impacts.*

Vulnerability is affected by development, so predicting and evaluating the significance of likely future scenarios demands the consideration of socio-economic changes and climate changes (Lorenzoni *et al.*, 2000). The Association of Small Island States (AOSIS) is concerned for the fate of low-lying Pacific islands, given that sea-levels will probably rise, and as oceans warm and bring new areas to the 26°C threshold, that may also mean wider occurrence of tropical cyclones (hurricanes and typhoons). Climatologists have become aware that ENSO events are a significant feature in the variability of tropical weather (and possibly that beyond the tropics). Interest in the phenomenon has increased since the event in 1997/98, which had a severe impact on many countries in South and Central America and further afield, including the USA where it is estimated to have cost billions of US$ (Chagnon, 2000).

4.7 Extreme weather

Historical records show occasional extreme weather events, rare enough for people to greet each occurrence as a 'surprise'. Often, especially in the tropics and sub-tropics, severe storms are associated with cyclonic activity; global greenhouse warming might make these storms more of a threat, extend the problem to new areas and possibly reduce risk for some regions (Longshore, 2000). The impact of successive similar severe weather events may each be very different for various reasons. In developed countries, citizens are more likely to have insurance and to make claims than would have been the case before the 1950s; consequently the cost of recent storms, floods or whatever is higher than it was in the past, even if the event is no more severe. Modern materials often make buildings more costly to repair and possibly more vulnerable to disasters. Plastic or aluminium cladding and roofing are vulnerable and can blow around, causing damage. There is less opportunity for the householder to make simple repairs than would have been the case in the past when small window panes were easy to replace, and new woodwork or thatch repairs could be commissioned locally or done by the householder at little cost. There has been a trend toward more easily damaged fittings and possessions. A traditional English country cottage hit by a sudden flood could be dried out afterwards with little expense other than repainting with a coat of whitewash; the furniture would have been solid wood, easily shifted to higher ground or an upper floor with little damage, and resistant to quite a bit of damp. The same dwelling today would have fitted carpets, difficult to move and easily damaged compression-board and veneer furniture, and vulnerable electrical wiring and appliances. So, almost as soon as the flood water arrives irreparable damage is done, costing a lot to remedy.

Before the 1940s few people in today's developed countries did much travelling to work or to buy supplies: employment was usually within walking distance or accessible by local train or bus, with the shops more than likely at the end of the street. Domestic heating was by coal, usually delivered in bulk during autumn, and there was limited dependence on electricity (if need be wood could be found and burnt in the fireplaces every home had); now many homes have no fireplace and fuelwood may be difficult to find in times of emergency. A heavy snowfall caused relatively minor disruption 50 years ago, whereas today people expect to drive miles to work and to shop, and often have no back-up for electric heating pumps, so chaos can be caused by severe weather. During the Second World War the population of the UK proved quite adaptable: strict rationing was enforced quite effectively, alternative power sources were developed for the relatively simple motor vehicles, and people proved willing and able to invest labour in growing food on unused land and in intensifying farming. One wonders whether modern citizens would be quite as adaptive and stoic. What is clear is that effective propaganda can play an important role in improving adaptation.

4.8 Failure to learn from the past

Simmons (1997: 50) noted that culture is a set of adaptations to the natural environment, which ensure the survival of the species. The widespread maladaptation of the present day might be because there has been a 'cultural lag' – people are still running on attitudes acquired during the long span of time in the past when their ancestors were hunter-gatherers or small farmers. The failure to respond to deteriorating conditions could be due to the 'boiled frog syndrome' (dropped in a pan of very hot water the frog would jump clear; placed in warm water and gently heated, the response might well be to stay put until it was too late): people have ignored gradually increasing problems where they would probably have responded if things had happened fast (Simmons, 1997: 50).

In a recent study of the political economy of hunger, Mike Davis (2000) argued that the last time the world eased trade barriers in favour of free trade it was followed by numerous famines in Ethiopia, Brazil and India. If Davis is correct, between the 1870s and early twentieth century 'liberal capitalism' helped create and sustain famines 'triggered' by climatic variations. Nowadays, global warming is expected to cause increasing climatic instability *and* there is a move to free trade again; things could be grim for the developing countries, if not the world as a whole, and in all probability the weather alone will be blamed again. Globalization of trade might have increased vulnerability (unlike Davis, I use the word 'might', rather than 'will' because market integration has provided some groups with opportunities for crop diversification, and thus helped make their livelihoods more secure against drought and pestilence). Yes, some will become more vulnerable to climatic or socio-economic factors by shifting from traditional subsistence livelihoods to cash-crop production (especially if this is monocropping), but increased vulnerability may not follow automatically.

There are economists who argue that globalization is making the world economy more vulnerable, so that a serious natural disaster in Japan, the USA or some other major trading nation would have more than national impact and could seriously upset world manufacturing, food and commodity production, and trade.

One must ask whether modern food production has learnt from past catastrophes. The 1930s Dust Bowl disaster in the US Midwest has nearly been repeated a couple of times since. The lessons should have been clear: manage the land better, control contract farming, make appropriate responses to drought and soil degradation. In 1845 the Irish Potato Famine was initiated by unfavourable weather and the arrival of a crop disease. It struck a large population relying on mono-cropping a food plant with little genetic diversity; poor governance, a repressed and pauper peasantry, exploitative landowners and the arrival of human epidemic diseases sealed the fate of perhaps a quarter of the island's population. The lessons the UK should have been learnt were: diversify food sources, reform land ownership and be prepared to give adequate emergency aid in time. Nowadays, too much of the world's agriculture is shifting towards monoculture, contract farming is spreading, and farmers who are keen to maintain profits are often reluctant to spend on better land husbandry.

Since the 1950s numerous large dams have been built around the world, many of these have generated very similar problems: resettlement, pollution, wildlife losses, disruption downstream damaging fisheries and floodplain resources, and seismic hazards. Large sums of money have been wasted through these problems, and many specialists have been employed in assessing the causes and finding ways of avoiding them. The striking lesson is that in spite of considerable hindsight experience, modern risk-assessment tools and vast expenditure, the same mistakes recur. Efforts to solve and avoid environmental problems are too often ineffective advocacy, and learning from the past seems a slow and haphazard process.

In many parts of the higher to mid-latitudes widespread formation and modification of soils occurred during and just after last glacial. The glaciers churned up the land and ground down rocks, leaving boulder clays, outwash deposits and areas covered by fine, wind-blown particles (loess). Throughout the last 10,000 years many of these soils have been demineralizing (leached of nutrients by the rain), and forming drainage-inhibiting layers. This has tended to mean gradually less fertile and progressively more acidic soils, even without the intervention of humans. One of the few who has paid much attention to this is Hamaker (1982), who argued that not only is the trend from high to low pH bad for agriculture, natural vegetation and humans, it might also reduce the amount of carbon dioxide being locked up by plants and thus trigger more global warming. He also claimed that demineralization somehow makes forest fires and bushfires more likely; the cure, he suggested, is to practise a form of organic farming and to dress soil with finely ground-up gravel or rocks of suitable type. Whether or not Hamaker is correct about natural demineralization, there is clear evidence of acidification of many farmed soils and some beneath natural vegetation. It is a problem that has grown since the 1750s, partly as a result of humans burning fossil fuels with a high sulphur content, and since the 1930s due to increasing use of chemical fertilizers.

Added to soil acidification are the threats of compaction (where soils are worked for crops or trampled by livestock), and pollution by industrial activity and agrochemicals. Humans are increasingly vulnerable to soil degradation – without fertile soil, food production would be impossible. The world expenditure on controlling soil degradation is meagre, and in developed countries like the UK those now educated in the environmental sciences often have little exposure to soil studies.

Key reading

Adams, W.M. (1992) *Wasting the Rain: rivers, people and planning in Africa*. Earthscan, London.

Blaikie, P.M., Cannon, T., Davis, I. and Wisner, B. (eds) (1994) *At Risk: natural hazards, people's vulnerability and disasters: El Niño famines and the making of the Third World*. Routledge, London.

Davis, M. (2000) *Late Victorian Holocausts*. Verso, London.

Simmons, I.G. (1997) *Humanity and Environment: a cultural ecology*. Addison Wesley Longman, Harlow.

|5|

Ongoing natural threats

Like other sciences, ecology has, in the last few decades, abandoned the 'machine models' that so long encouraged its practitioners to imagine in nature order and simple predictability ... in favour of a new emphasis on the complexity of natural systems.

(Athanasiou, 1996: 260)

Chapter summary

This chapter reviews ongoing natural threats, some terrestrial, others of extraterrestrial origin. The previous chapter explored present vulnerability; this chapter notes the lack of awareness, and limited willingness of governments and people to pay sufficient attention to threats.

The years 1990–2000 were the UN International Decade for Natural Disaster Reduction. (Promoted by the US National Academy of Sciences, it has established better links between scientists, engineers and decision-makers, and 150 nations participated.) Its goal was to create and maintain a safer environment by reducing the loss of life and damage caused by natural disasters worldwide. Although there have been advances in the fields of hazard and risk assessment, and disaster relief planning, the avoidance and mitigation of natural threats depends very much on the willingness and ability of governments to plan ahead, and their ability to react swiftly in an appropriate way. There will doubtless be debate in the UK about the contingency planning and responses to foot and mouth disease of livestock in 2001/02 – a relatively fast-onset hazard. Given that outbreaks have affected the UK in the past, with major occurrences as recently as the late 1960s, the less than well-orchestrated responses during the 2001/02 outbreaks indicate that the UK is no less vulnerable than it was in the past.

Known threats that have a slow onset should give more time for response, unless they develop so gradually that people fail to perceive the situation until it is too late. Unknown threats, whether fast or slow to develop, pose challenges; at least when the onset is slow there is a chance to realize the nature of the problem and perhaps conduct research.

There are three other important components to factor in to any attempt to calculate the seriousness of a threat. These are:

- extent
- frequency
- severity.

Where known threats recur only occasionally people may assume it is not a problem for their generation. For example, a once-in-200-years flood risk; a once-in-a-thousand-years tsunami; comet or asteroid strikes every million years or so. The problem is that these threats may be rare, but can be catastrophic. Earth history shows that comet or asteroid strikes can wreck continents, have probably come close to destroying all life at least once and have repeatedly caused severe disruption. However, it is difficult to prompt investment in the avoidance or mitigation of something that, no matter how many it may injure or kill, is expected to happen only every several thousand years, or less frequently. Ironically, humans – the first of Earth's organisms in over 4500 million years able to perceive threats and with the technology to do much about them – may be too short-sighted to act.

One might divide continuing threats into 'earthly' (terrestrial) and 'exogenous' (extraterrestrial). Various bodies, agencies and individuals do publish warnings about environmental threats, and sometimes monitor the situation. Usually the focus is on specific hazards and problems, not the overall picture. Attempts to assess the range of challenges and to suggest priorities are less common and, when done, are likely to be superficial and fail to go beyond crude advocacy. Concern also tends to follow 'fashions': in the 1930s, thanks to the Dust Bowl disaster in the US Midwest, there was for a while strong interest in fighting soil degradation. From the 1960s until the 1980s concern was voiced about human population growth, pesticide pollution, the risk of atomic warfare, careless technology and the need to improve food production. The 1968–73 Sahelian drought prompted worries about desertification. Between the 1980s and 2000 attention focused on global warming (in contrast to the 1960s and 1970s when warnings about future cooling were often published), loss of biodiversity and stratospheric ozone depletion. In the 1990s the West began to fear biotechnology, especially genetic manipulation (GM) and carcinogenic compounds.

Reviews of natural threats include: the 'State of the World' reports from the Worldwatch Institute (Washington, DC), which have appeared annually during the past few decades (Brown *et al.*, 2001); occasional publications from international bodies like the UN and the World Bank, notably *The World Development Report 1992* (World Bank, 1992) and *The Global 2000 Report to the President* (Council on Environmental Quality and the

Department of State, 1982). A number of publications have had some success in raising awareness that the environment has limits, that environmental issues demand serious attention, and that an anticipatory or precautionary (forward-looking) approach is desirable; these include: *The Limits to Growth* (Meadows *et al*, 1972); *The Next 200 Years* (Kahn *et al.*, 1976); *Our Common Future* (World Commission on Environment and Development, 1987); *Beyond the Limits* (Meadows *et al.*, 1992); and the writings of James Lovelock on the Gaia Hypothesis (Lovelock, 1979; 1988).

There is a certain mindset amongst the specialists who advise governments. As Ager (1993: xi) noted, 'For a century and a half the geological world has been brain-washed, by the gradualistic uniformitarianism of Charles Lyell. Any suggestion of "catastrophic" events has been rejected as old-fashioned, unscientific and even laughable.' The history of the world is not just of gradual and steady change, but also of sudden, perhaps seemingly 'improbable', catastrophes, and episodic or occasional events. Planners and administrators have to cover both, using sensitive long-term monitoring for tracking subtle changes, and identifying, assessing and making contingency plans, as well as taking quick evasive action against unexpected 'moments of terror'.

Unfortunately, the recent past and present may not be very representative when one is interested in the full range of possible catastrophes. Even for relatively 'frequent' events (on a geological timescale) history is too short a span, although mythology may hint at some longer memories. Infrequent but catastrophic events do deserve attention but concern for the 'mega' must not swamp interest in the lesser-magnitude threats that are more frequent. In cricket a batsman uses a restricted range of responses to defend against most bowls, but success demands ability to react appropriately to sudden unexpected threats and opportunities.

The significance of a threat depends on how and when humans are affected, and particularly upon their vulnerability at the time it materializes. It is also possible for more than one environmental threat to act cumulatively. Identifying these cumulative, and often indirect, threats is difficult. The following sections review ongoing challenges.

5.1 Food shortage

Interruption of food supplies is one of the key ongoing threats. The historical causes have been 'earthly' (bad weather, drought, warfare, insect pests, fungal diseases, natural soil degradation and so on), but could conceivably also be the result of extraterrestrial causes such as increased solar radiation or comet strike, or gamma-ray burst. History has shown that crop failures have often been followed by disease epidemics and longer-term socio-economic problems. In the past, before the twentieth century, most people worldwide were in a 'tight balance' between energy expenditure on food production (human labour and draught animals) and the energy their labours generated (food and fodder); poor weather or other misfortune easily led to hunger. In some societies still that situation prevails.

Food reserves 'peaked' in the 1980s, and since then there has been a fall in grain surpluses worldwide. In 1992 there were probably only grain reserves to provide the world's demand for 62 days. Between 1958 and 1961 the People's Republic of China had severe famine, and again in the early 1970s (the latter, some say, killed as many as 40 million; a figure of over 14 million seems likely), and North Korea is in the grip of severe food shortage at the time of writing (2002). People today see threats like locust swarms as a thing of the past. Yet, southern Africa was severely hit in 1987 and present shortages of funding, reluctance to use environmentally damaging insecticides, plus warfare and unrest that prevents monitoring of breeding areas, means the problem could again flare up. Reports in *The Times* (UK, 19 June 2001: 11) indicate growing locust problems in Kazakhstan, China, Africa and even the USA. The insidious problem of soil degradation has already been discussed, yet relatively little effort and funding is being devoted to combating this. Soil, is of course, crucial to food production and often virtually impossible to rehabilitate or replace once it is degraded. More attention must be directed at improving and sustaining food production. Also, if current grain surpluses are not increased there is little chance that reserves will be improved enough to be available as a buffer against short-term disruptions of harvests.

5.2 Earthquakes

Earthquakes pose a direct and indirect threat, in that they can cause fires, tsunamis and landslides as well as shock damage. Earthquakes are among the main killers of humans, with some claiming hundreds of thousands of lives. Most, but not all, seismic activity is along the edges of continental plates, the risk is present everywhere, although occurrences in some areas may be very rare. A region may be aware that 'quakes pose a recurrent threat, but actual strikes occur suddenly and often with little obvious warning.

Recent studies suggest that even minor local tremors or distant earthquakes might cause damage where settlement is located above geological structures that act as a 'magnifying lens' to concentrate and focus seismic waves. This may have been the case with Kobe (Japan) in 1993 (when over 5000 were killed), at Northridge (California) in 1995 and Santa Monica (California) in 1996 (O'Hanlon, 2001). A seismic lens may be formed by a boundary between different rocks; shock waves pass faster through one rock than another and are deflected. If the effect can be proven and understood seismologists might well be able to identify areas at risk using techniques familiar to oil prospectors (gravimetric and seismic reflection studies).

5.3 Tsunamis

Tsunamis are waves of greater than normal height and wavelength, which sometimes sweep across oceans and other water bodies. Although nothing to do with tides they are commonly called 'tidal waves.' Humans have long feared tsunamis; myths, for example, tell of the sudden inundation of the civilization of Atlantis. What prompted the tale is uncertain; although, among many often wild theories, plausible possibilities include the following.

- The explosive eruption of Santorini around 1628 BC. The crater (caldera) lies where the Aegean island of Thera is now sited. This eruption possibly destroyed the Minoan civilization on Crete and elsewhere in the eastern Mediterranean with ash falls, dust-veil cooling and acid deposition as well as tsunamis. On Crete the tsunamis left pumice boulders 250 metres above present sea-level. The pre-eruption shape of Thera appears to match descriptions of the distinctive layout of Atlantis.
- The legend may be prompted by the loss of the city of Helike, probably sited on the Gulf of Corinth, to tsunami and earthquake (around 373 BC).
- That Atlantis lay in the western approaches to the Straits of Gibraltar.

The threat of tsunamis is understated. Even in recent decades in Japan alone there have been 20,000 deaths due to relatively small tsunamis. More and more development takes place close to coasts, including the siting of nuclear power stations. Throughout the tropics the loss of sheltering mangrove thickets and coral reefs has increased vulnerability. California, Florida, Papua New Guinea and many other regions have offshore continental shelf ravines that can suffer underwater mudslides that generate serious tsunamis.

Some sub-divide tsunamis into 'regular' and 'mega-' (or 'super-'). The former may reach 40 metres or so above sea-level when they come onshore. Mega-tsunamis are known from palaeoecological evidence and have exceeded 300 metres' reach. Tsunamis cross oceans as very long-wavelength pulses that cause terrible damage when they reach shallows. The Storegga slide (see Chapter 3) of about 7200 BP may have helped separate the UK from the continent and flooded low-lying lands across northwestern Europe, but was probably less than 20 metres high.

Regular tsunamis vary in height according to the effectiveness of the trigger event, the distance and conditions under which the waves travel, and the slope of the shore and shape of the coastline struck. When a tsunami reaches shallows its leading edge slows and the following water piles up, causing the wave to rear up. More than one tsunami may be generated; in the case of the Krakatoa eruption in AD 1878 several waves struck along the Sunda Straits to over 40 m above normal sea-level, killing over 36,000 people. In 1998 Papua New Guinea suffered a pulse of three *c.* 15-metre tsunamis, which killed thousands (Bryant, 2002). The Lisbon earthquake of

AD 1755 generated several tsunamis of around 5 metres above normal sea-level. The earthquake, widespread fire and tsunamis killed around 60,000 of Lisbon's population.

Bays, inlets and even lakes well inland, like those of Switzerland, can suffer huge waves; one example is Lituya Bay (Alaska) where, in 1958, an ice fall from a glacier generated a tsunami over 60 metres in height, which stripped trees and soil from slopes 490 metres above sea-level – illustrating the way in which a moderate-height wave can rear up to a great height. In 1965 the Vaiont dam (Italy) was overtopped by a wave generated by a landslide; it failed and killed many people down-valley. In the 1990s a small boat was carried by one enormous wave in an Alaskan inlet and its crew members survived. That wave, which stripped soil and trees away as high as 525 metres above sea-level, was triggered by a landslide. Enclosed bodies of water may also experience 'seiches' (rhythmic movements, typically forming standing waves as water oscillates back and forth because of wind action or seismic activity). One might speculate that the Biblical story of the escape by Moses and the Israelites from pursuing Egyptian forces across the Red Sea might have roots in a seich where water levels repeatedly rose and fell, or a tsunami where the sea withdrew for perhaps as much as half an hour before the wave struck (a well-established phenomenon).

Countries like Japan, Hawaii and Chile suffer tsunamis sufficiently often for it to be worth their while installing special breakwaters, controlling land use, and developing early warning, public education and evacuation procedures. Early warning is now possible using US NOAA (National Oceanic and Atmospheric Administration) satellites, which relay data from wave-measuring devices: it takes *c.* 15 hours for a tsunami to travel from Hawaii to Chile, and *c.* 5 to San Francisco – so several hours' warning is often possible. Hawaii publishes evacuation advice in phone booths to try and ensure that people know how to react when warned.

Mega-tsunamis have not happened in historical times, but there is convincing geological evidence that they are reasonably common, and this is reinforced by myths in northwestern North America and elsewhere. Hawaii may have experienced such a wave roughly 2 million years ago (caused by the Nuuanu landslide) when huge blocks of rock from Oahu Island were dislodged into the sea during an earthquake or eruption – it is believed to have hit Hawaii, the Pacific Rim (including South America), western USA and China. This tsunami deposited debris on other Hawaiian islands as much as 375 metres above normal sea-level. Ocean floor surveys have identified at least 17 similar huge landslide sites around the Hawaiian islands.

The Cape Verde Islands and Canary Islands seem capable of generating mega-tsunamis, and probably other steeply sloping oceanic islands like Tristan da Cunha, Réunion and the Marquesas generated similar waves. Studies suggest that huge waves from Atlantic islands could easily strike Europe, eastern USA and the Caribbean. There are signs that giant waves generated either by huge tsunamis or (less likely) tremendous storms carried large boulders to 20 metres above sea level, and probably submerged much of the Bahamas about 120,000 BP, this seems to coincide with a huge landslide on El Hierro (Canary Islands) (Marshall, 2001). Réunion may have caused a tsunami about 4000 BP, which hit Australia.

In the future volcanic activity on La Palma (Canary Islands) could generate a mega-tsunami, which would cause tremendous destruction in the eastern USA and the Caribbean. The Taburiente volcano seems to have done so in the past, but is probably extinct. However, Cumbre Vieja in the south of the island is still active, and studies suggest that its flanks could split during seismic activity and generate huge waves – it has erupted twice in the last 200 years or so – most recently a few decades ago, causing some worrying faultlines and subsidence. Risk assessments indicate that Cumbre Vieja's slopes are highly unstable and that mega-tsunamis might be caused during the next 100 years. Effective monitoring, which spotted the start of a mega-tsunami, could give the New World about eight hours' early warning. Before a landslide happens vulnerable installations like nuclear waste repositories, important libraries or gene banks should be relocated. (For those wishing to read further on this subject, there are a number of journals and websites that deal with tsunami hazard, notably the *International Tsunami Society* at http://www.ccalmr.ogi.edu/STH/current.html, accessed July 2001.)

5.4 Landslips, mudslides and avalanches

In most highland regions people have traditionally sited settlements and infrastructure to minimize landslide, mudslide and avalanche risk. Nevertheless, vulnerability can change: glaciers may expand, as was the case during the Little Ice Age, beyond anything memories warned of. In recent years hillside trees have been damaged or killed in many uplands by over-grazing, woodcutting, bushfires, air pollution, acid deposition and (in richer, higher-latitude nations) skiing. The damage reduces the protection from avalanche and rockslides enjoyed downslope and often accompanies increased settlement because of tourism development. In some regions the shortage of suitable land has driven the poor to build their shanty towns where they are vulnerable; sometimes the rich court similar risks in order to have homes with spectacular views.

The threats in steep areas are generally predictable and may be mapped. Tree planting and engineered structures can be installed to reduce the risks (if there are funds), and regulations can be used to prevent building or activities in dangerous locations. Mass movements are more likely if there are earthquakes, storms or bushfires, and authorities should be extra vigilant during such events, wet seasons or El Niños. However, where a threat has increased to endanger areas already settled or where there is not the money or inclination to take preventative action there will be problems.

Avalanches, landslides and mudslides can cause serious loss of life and damage to infrastructure; for example, over 75,000 died in northern Peru alone in 1970.

Earth movements can have indirect impacts: in parts of the USA, notably the Los Angeles area, landslides have caused outbreaks of coccidioidomycosis ('valley fever'), by liberating spores that become airborne to infect people through inhalation. Similar disease outbreaks may occur elsewhere, going largely undetected where there is less developed healthcare. Landslides can also dam rivers, causing flood risk or flow-diversion problems.

5.5 Volcanic activity

Over the last hundred years or so volcanic activity has killed roughly 82,000 people, with about 17 eruptions out of several hundred accounting for the bulk of those fatalities. While there is a chance that an eruption can take place where there has been no previous activity, and with no obvious warning, the problem is mainly confined to known and mapped regions, where crustal processes make vulcanicity far more likely. Active areas might shift or new areas form, triggered by events like asteroid strikes. In the past there seem to have been long periods where vulcanicity was far more frequent and serious than today (for example, in the Miocene, 12–14 million years BP); Courtillot (2000) suggested that at least seven of the Earth's known mass extinction events may have been related to vulcanicity.

Volcanoes mainly appear in two locations.

- In belts inland of coastal zones of subduction where the sea floor slowly plunges below a continental plate. At depth, the sinking rock becomes heated, plumes of molten rock and gases rise from 'hotspots' breaking through at the surface as volcanoes.
- Along mid-ocean ridges where the sea floor is splitting and spreading apart there are also likely to be volcanoes. However, it is also possible to have volcanoes elsewhere, even in mid-continental areas, especially if the crust has been weakened by faulting, extraterrestrial body impacts or rising hot convection currents from the magma.

Worldwide, there are now about 550 active land volcanoes, mainly clustered along plate margins. The question is, might very large shield volcanoes (like the Deccan Trap outpourings) happen in the future and, if so, where, and when? There is a more concrete risk from massive explosive caldera-type eruptions; although rare, these do happen frequently enough to be a serious threat, and could cause catastrophic regional blast damage, perhaps tsunamis, and serious global climate impact. In Europe the known risks are close to the populous areas of Naples and Capri – the volcanic areas of Vesuvius and the Flegreian Fields. Sicily may also be at risk. In

Indonesia there are risk areas and in the USA, the Yellowstone National Park, the Cascades and Alaska pose a serious threat. The Yellowstone region has had massive eruptions within the last million years, and another may be 'overdue'. Earth scientists have been checking for crustal deformation that might indicate a rising magma plume long enough before Yellowstone explodes to allow some warning. There are many other potential disasters, including Mount Rainier (USA), which threatens around 100,000 people, and in the Philippines, Mount Pinatubo is surrounded by dense settlement.

Volcanoes are often quiescent for centuries and then suddenly erupt, sometimes violently; however, there has been some progress in monitoring those that are dormant and suspected to pose a threat. The art is far from perfected but might give enough warning for evacuation, though there have been a few costly false alarms. There is a possibility that, *if* volcanic activity is found to be periodically more likely, this may aid forecasting. Monitoring tries to spot the signs of a rising mass of hot magma before it actually erupts, using geomagnetic surveillance, ground deformation measurements, resistivity surveys, ground temperature checks, analysis of gas emissions, and local seismic activity observations. Some recent forecasts based on monitoring seismic activity have proved quite accurate (Marshall, 2002).

Because volcanic activity can be explosive or involve fissure-type outpourings, it presents a number of different threats, as outlined below.

Blast damage, seismic shocks and tsunamis

Tambora (Sumbawa Island, Indonesia, 1815), the largest eruption of the last 10,000 years or so, killed over 12,000, and Krakatoa (Indonesia, 1883) 36,000 or more. Even during the last decade there have been quite a number of eruptions that have had serious regional impacts: Mount Pinatubo (Philippines, 1991), in spite of quite careful monitoring and warning systems, killed 700 people, displaced 500,000, damaged and closed a US airbase, and probably eliminated around 40,000 jobs.

Fall-out

The material ejected by volcanoes ranges from large 'bombs' (golf-ball size to metres in diameter), to lapilli (roughly pea-sized), and coarse or fine ash. The debris may be very hot on return to Earth – a pyroclastic fall – and may ignite vegetation and buildings. Larger particles pose a threat because they remain hot longer and are heavy enough to injure organisms and damage property. In violent eruptions, ash may be mixed with a variety of volcanic bombs. The effect is to hinder or prevent evacuation, choke and bury crops, buildings and people, and contaminate water supplies.

Typically, there is a depositional pattern determined by the wind, and the height the material is ejected to, with smaller particles drifting further from the volcano and larger settling close in. During an eruption the size and chemical character of the ash and other debris ejected may change. Ashfall may take place miles downwind of an eruption in sufficient quantity to collapse buildings, halt road and other communications (including radio, due to the static discharges), cause respiratory distress, ruin water supplies, and damage vegetation, soil and crops. Once weathered, old ashfalls may form quite fertile soil, but not always; some ash, pumice and lava may be contaminated with poisons or be too free-draining or solid and impenetrable for revegetation, and very slow to break down into soil. Sometimes desolate wastelands are created, which are difficult to cross on foot or by vehicle, and which show little sign of allowing vegetation recovery centuries later.

In Japan, Mount Sakurajima commonly ejects ash, so local people have built concrete shelters. Not all volcanoes are as predictable, and it may not be wise to trust any. The risk is that ashfalls may drive people to shelter in buildings rather than evacuate, and they are then vulnerable to lahars and pyroclastic flows, which is what seems to have happened when Vesuvius erupted in AD 79 (Fig. 5.1). High-altitude ash clouds present a serious danger to aircraft because they can stop jet engines.

Figure 5.1 A main street in Pompeii, buried under volcanic ash and pyroclastic flows in AD 79
An environmental disaster unequivocally correlated with past human misfortunes.
Source: Author 2000

There are bodies that have started to offer ash cloud warnings to aviation – for example, the Smithsonian National Museum of Natural History, Global Volcanism Program (provided by NOAA, Satellite Analysis Branch). There are also volcanic activity reports issued by various bodies around the world. In future monitoring and warning is likely to improve.

The solidified ash deposits (which can be several metres or more thick, even scores of miles from a volcano) are known as tephra. Often, individual eruptions can be recognized by the characteristic ash particles blown around the world, offering palaeoecologists correlation and dating horizons.

Lava flows

These can be fluid and fast moving, or viscous and slower. The latter type can block a vent and cause violent explosions. Lava flows in recent geological history have usually been localized and slow enough for people to escape them (Fig. 5.2), but there have been huge outpourings in the past. The eruption of Mount Nyiragongo (Democratic Republic of the Congo, 2002) caused roughly 300,000 to flee the town of Goma when rapid lava flows escaped from a lava lake in the crater at speeds of up to 70 kmph. Some volcanoes are quite variable in output, alternating between lava flows,

Figure 5.2 A house engulfed by a lava flow on the northwestern flank of Etna (Sicily)
The occupants took refuge on the roof and were evacuated unharmed. The highway was simply rebuilt.
Source: Author

ash emissions and gas releases; others are more constant, and a few are unpredictable.

Mudflows (lahars)

Fatality numbers due to lahars are quite high, compared with other volcanic phenomena (Mastin and Witter, 2000). Lahars are mudflows associated with volcanoes with crater lakes or snow caps, but can also be generated by intense rainfall on to ash-covered slopes. They travel fast, often far from the volcano, and may suddenly inundate valleys where there is often dense settlement.

Mount St Helens (1982) claimed 57 victims, most through lahars. In 1984 the Colombian volcano Nevado del Ruiz melted an ice cap at over 5000 metres' altitude, lahars swept down valleys, killing 23,000 people in the city of Armero alone. The volcano has a history of causing such disasters, and warnings to evacuate were issued about eight hours before the lahars struck. Unfortunately people chose to ignore the warnings.

A phenomenon similar to a lahar is a glacier outburst, generated in more level terrain by vulcanicity under a glacier or ice cap. Iceland suffers these flash floods, which threaten along a wide front and carry glacial debris including large boulders. An Icelandic term is applied to this phenomenon: jökulhlaup. A particularly large one occurred at Skeidararsandur in 1993.

Pyroclastic flows

Hot ash and gases can blast from a volcano as an incandescent cloud (nuée ardent) or, driven by gravity and lubricated by the expansion of gases entrapped in the frothy lava, avalanche considerable distances. These pyroclastic flows can suddenly punctuate an otherwise fairly gentle eruption, sweeping at speeds of over 100 kmph, searing everything in their path and doing considerable blast damage perhaps tens of miles from the volcano: sometimes they jet out sideways from a volcano and can ride upslope over ridges and hills, and cross lakes and stretches of sea. Only those sheltering well below ground level and in very solid buildings have a chance of survival, and shipping within a few miles of shore is also vulnerable. Early in the twentieth century Mont Pelée released pyroclastic flows on to the capital of Martinique, St Pierre; around 30,000 died in a few moments, and there were only two survivors (Zebrowski, 2001).

Aerosol and gas emissions

If an explosive eruption blasts dust and acid-rich aerosols into the upper atmosphere it can spread on the jetstreams. The ash and sulphuric acid-rich aerosols thrown up by the Pinatubo eruption were tracked around the world and appear to have disrupted the climate for five years or more, depressing global temperatures. The likelihood of this depends on the size of the blast and possibly the location, with high-latitude sites more likely to contaminate the upper atmosphere. The dust veil reduces penetration of sunlight and causes acid deposition, which injures plants and animals, and might disrupt stratospheric ozone. The explosive Tambora eruption of AD 1815 was the greatest ash eruption of the last 10,000 years, and is blamed for global cooling (a reduction of Northern Hemisphere mean temperatures by 0.4° to 0.7°C for a few years). The damage seems to have been done by sulphur dioxide and sulphide emissions rather than ash alone, and these stayed high enough to chill the globe for about two years.

Some relatively small fissure eruptions also seem to alter climate and cause widespread acid fall-out, although the mechanisms by which the gases, dust and aerosols are dispersed (possibly as a 'dry fog') are less clear. The eruption of Laki (Iceland) in AD 1783 probably did not blast material very high, but did emit a lot of sulphuric compounds, and is blamed for bad summers in Scandinavia and Iceland, and for poor harvests and faltering economies in Europe for some years after. Both fissure and explosive eruptions seem capable of generating clouds of sulphur dioxide, hydrogen sulphide or carbon dioxide ('dry fogs'), which may drift over quite distant regions and have ill-effects on people, animals and plants.

Even the small fissure eruptions of recent times in Iceland and Hawaii have necessitated the evacuation of surrounding areas, and agriculture has been ruined at considerable distances. Truly frightening fissure eruptions (flood basalts or traps) were erupted before humans evolved: the Columbia River basalts (northwest USA) were deposited 40 to 60 million years ago, cover 200,000 km^2 and average more than a kilometre in thickness. One can imagine the impact such vulcanicity would have on the society and economy of a continent today, and the effects it would probably have on global climate and atmospheric composition. What is clear is that these outpourings have recurred through geological time. What is less certain is the cause.

Not all volcanic gas releases take place during eruptions; carbon dioxide and other gases may seep from volcanic areas to accumulate in caves or hollows, or in solution at the bottom of a lake or in the sea. Invisible pools of gas in caves or hollows can kill animals when they bend to drink or eat; scavengers are attracted and suffer a similar fate, so the chances of preservation and fossilization are much enhanced, providing palaeontologists with some of their best finds. In recent times there have been sudden gas releases from volcanic crater lakes, which have killed wildlife, livestock and people in the surrounding area. An outgassing from Lake Nyos (northwest Cameroon) killed at least 1746 people in 1986. A similar outgassing happened at Lake Monoun (Cameroon) in 1984 when 34 people died, some as far as 23 km from the Lake. Recent studies suggest that the Camroonian lakes were stratified because of little daily or seasonal temperature change, insufficient streamflow entering, or wind activity. Stratification means that lake waters fail to mix and gases stay in solution at depth; an unusual temperature change, a landslide into the lake or an earthquake has a similar effect as the removal of a cork from a champagne bottle. Where the threat is identified and funds permit, it might be possible to de-gas a lake in a controlled manner so that its unexpected outbursts are tamed (Jones, 2001b).

5.6 Oceanic outgassing

Gases like methane, hydrogen sulphide, carbon dioxide and sulphur dioxide generated by decomposition of organic matter or escaping from the Earth's crust can accumulate in sea- or lake-bed sediments, sometimes as pure, solid deposits (methane and possibly other gases solidify, when the pressure and temperature allow, to form solid gas hydrates or clathrates), or are held in solution. Quite shallow marine clathrate accumulations are known at high latitudes; they occur at greater depths at lower latitudes, and storage may also take place in deep lakes and cold peat or soils. If warmed or subjected to a pressure reduction stored gases can suddenly bubble to the surface (oil-rigs off Alaska and in the North Sea have set off quite large outgassings). Huge outgassings might well be triggered by seismic activity, mudslides, altered ocean currents and global warming; what is not established is whether escapes could alter global climate through carbon enrichment or have onshore impacts. Outgassings might explain climatic shifts and ice core evidence of atmospheric carbon increases that do not correlate with volcanic eruptions. There has also been speculation that clouds of methane and gases like hydrogen sulphide might drift over land, poison vegetation and make animals and humans ill, or ignite over sea or land causing fire storms.

Outgassing might sink passing ships by lowering their buoyancy and affect aircraft in the vicinity. Large blemishes in the floor of the North Sea may record such outgassings (one is reported to have the wreck of a trawler lying relatively undamaged in it). Tales about lost ships and aircraft in the 'Bermuda Triangle' might be founded on some such phenomenon. There are no reliable witnesses of an outgassing; although, a sea captain off the Azores in AD 1775 recorded what may have been a large one (possibly set off by shocks associated with the Lisbon earthquake). Bailie (in Slack, 1999: 60–2) sought to stimulate interest in the phenomenon, which he suggested might explain some past human epidemics blamed on plague. If global warming causes widespread release of methane it could act as a positive feedback, increasing global warming.

5.7 Drought

Simple shortage of precipitation may not always lead to significant drought impacts; drought can result from many causes, and varies in its impact

(Fig. 5.3). Drought is often seen as something affecting only drier regions. However, there can be periods when rains fail to supply enough moisture for plant growth and allow vegetation to dry out even in humid environments: the UK has had serious droughts as recently as 1975/76; in the 1980s the Panama Canal ran short of water to feed its locks; and rainforests in Amazonia, central America and Southeast Asia sometimes dry enough to suffer bushfires. Nowadays, many regions are undergoing increasing demands for available water supplies, so drought is a growing threat. Worse, the available supplies are often polluted. The administrators of a number of cities, like Mexico City, are concerned over how they will meet future urban water demands. During drought, dust may blow and settle at great distances, that from the Sahara and its surroundings reaching the Caribbean and leaving a record in mid-Atlantic Ocean sediment cores, that from China drifting as far out into the Pacific as Hawaii.

Drought struck southern and central Asia in 1972, hitting Soviet wheat-producing areas and forcing the USSR to purchase 28 million tonnes of grain from the USA and Canada, which significantly affected world food prices and reserves. Most of the world's main grain-exporting areas, including those of Russia, North America, Western Australia, central-south Brazil and Sicily are prone to occasional drought. The Sahel (sub-Saharan West Africa and the Atlantic Cape Verde Islands), Ethiopia and the Sudan suffered severe drought between 1968 and 1973. In the past there were smaller numbers of people in drought-prone areas and, provided they

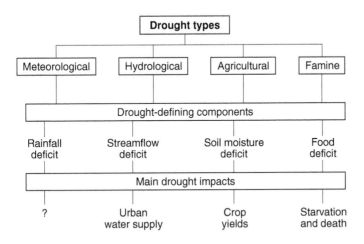

Figure 5.3 Types of drought
The potential for disaster increases from left to right (rainfall shortage alone may not lead to significant drought impacts).
Source: Reproduced with permission from Smith, K. (1992) *Environmental Hazards.* Routledge, London.

realized the threat in time and moved, there was often unoccupied land for them to relocate to. Today unoccupied land is scarce and movement may also be hindered. The threat of drought has also increased for some peoples because their drought-avoidance or mitigation strategies have degenerated or been lost.

Increasing evidence is being gathered about the linkages between phenomena like El Niño and drought around the world. Although not wholly reliable, this knowledge should give forecasters a greater opportunity to identify drought-risk regions many months before problems strike.

5.8 Floods

Flood and storm damage seems to have increased in developed countries. In part this is due to expansion of settlement on to flood-prone areas and increased cost because of easily damaged possessions that are more likely to be insured than in the past. If people settle a canyon or a floodplain created by flood water they should not be surprised when inundated, yet they often are! The year 2001 was memorable in the UK for the extensive, severe, repeated flooding; weather can change, although – partly attributable to altered land use, neglected upkeep of flood defences and recent building in flood-prone areas – the inundations were mainly because 2001 was a year of above-average precipitation.

In many parts of the world floods are tolerated or even welcomed; people and other organisms adapt and take advantage of the opportunities: Nile floods provided fertile soil-enriching silt and irrigated the land, especially before the Aswan Dam was built. Peoples in Amazonia practise seasonal transhumance as an adaption to flooding and, in Iraq, the Marsh Arabs are adapted to swamp environments that flood. Some peoples are poorly adapted to floods: Bangladesh regularly suffers storm surges and people have limited mitigation strategies; Mozambique is prone to extensive river floods, yet little has been done to adapt. There are hopes for better warnings once El Niño effects are better understood. Forecasting has already been improved through the adoption of weather radar, better sharing of flood data between nations and improved monitoring of cyclones (which can cause storm surges and heavy rainfall).

5.9 Severe weather events

Occasional severe weather events are inevitable in virtually all parts of the world. The main threats are: storms (especially cyclonic storms – typhoons and hurricanes); tornadoes; various other types of storms with heavy wind, rain or hail; extreme or unseasonable frosts; unusually wet or cold seasons; heatwaves. A cyclone is a storm system of high-speed winds that rotates about a centre of low atmospheric pressure, bringing heavy rain and usually following fairly predictable paths. A typhoon is a tropical cyclone, especially in the Philippines and South China Sea regions; a hurricane is a tropical cyclone of the Western Hemisphere, common in the Caribbean, Central America, northern Latin America and southern USA; hurricanes also occur in the Pacific. It is usual to name a hurricane that is being tracked, using a feminine or masculine first name (e.g. Freda). The strength of storms is usually measured on the Saffir-Simpson Scale of category 1 to 5; wind speeds are usually indicated with reference to the Beaufort Scale of 0–14, with Force 14 being a hurricane at over *c.* 150 kmph. A tornado is a type of whirlwind that can cause great damage along a narrow track. Certain regions are currently more prone to tornadoes than others because of their relief and location – notably the US Midwest. Tornadoes can also occasionally occur in Europe and in many other parts of the world. A tornado forming over the sea may touch down to become a water-spout, which may be a hazard to shipping.

Sometimes hurricanes and typhoons wander from their 'normal' tracks, and may do so more if global warming occurs (see Chapter 4 for discussion of a possible link between hurricane behaviour and El Niño events). The New York–New Jersey coast (USA) was hit by a severe (category 3) hurricane in 1821, and subsequently on three occasions by hurricanes that were less powerful, yet most would regard this area as beyond the usual range of hurricanes. Palaeoecologists have found evidence of what they think was a category 4 hurricane, which struck near Atlantic City (on the east coast of the USA) sometime between AD 1278 and 1438. Clearly, there is a history of occasional severe weather in regions now much more densely settled and with high-rise buildings that may be vulnerable. In theory, if global warming raises sea surface temperatures above 26°C there is more possibility of these storms being generated, although Burroughs (1997: 75) could find no evidence for a clear trend toward an increased occurrence over the usually affected Atlantic areas.

5.10 Climate change

Particularly since the 1750s humans have been modifying the natural environment. Consequently, it is now often difficult to separate out natural climate change and anthropogenic. Whether induced by humans or wholly natural, climate change is an ongoing thing, which may challenge people or offer them opportunities (an example being the recent warming that has led to the expansion of vineyards in the UK). Hardly a day goes by without the media reporting on global warming. The past 50 years were hailed as the warmest period of the millennium and glaciers are receding worldwide. The assumption is that the long-term trend will be toward warmer conditions, or at least temperatures similar to now, and efforts are focused on countering anthropogenic warming. It would be better to pay attention to the probability (if not certainty) that climate can suddenly change, whether this is driven by natural causes alone or influenced by human activities.

Many environmental historians argue that between the thirteenth and the late seventeenth century the Little Ice Age (see Chapter 3) placed humans under severe stress. The question is, was it a 'random blip' in a generally benign postglacial climate, or a periodic natural change (human activity may decrease or increase the chance of recurrence) (Pearce, 2000a). Some climatologists suspect there may be a more or less regular *c.* 1300–1800-year pulse in world's climate, whether or not there is a glacial period or warmer conditions prevailing. The evidence comes from temperature records in ice cores and ocean sediment cores. It is suggested that a *c.* 1500-year climatic pulse (not always a regular pulse) is caused by solar fluctuations and that some, possibly random events, coinciding with the right point in this fluctuation could trigger sudden warming. The palaeoclimatic record shows several of these sudden flips with as much as 10°C warming in less than a decade (known as Dansgaard-Oeschger warmings). So, a sudden shift could also be to warmer conditions, as well as cold. Clearly, it is unwise to assume that global warming will result in gentle and gradual warming in the foreseeable future.

Ocean sediment cores also reveal Heinrich events (see Chapter 3) – layers of iceberg-rafted rock fragments deposited roughly every 8000 years during the last ice age on the Atlantic floor as far south as Bermuda. Heinrich events reflect increased calving of icebergs from the Arctic and Canada, probably initiated by shifts in ocean circulation. In the Northern Hemisphere, the ice-rafted debris record from ocean sediment cores alternates from carbonate-rich (during times of extensive ice) to quartz-rich (indicating altered ocean circulation) when conditions warmed (Bischof, 2000).

Pearce (2001) reported research on 'desert varnish' (distinctive coloured encrustations on rocks in arid environments), which offers a means of establishing palaeoecology, dating and correlation of changes on land with Heinrich events. Ocean sediment core plankton records, and growing evidence of past temperature and rainfall shifts in Amazonia and tropical Africa, correlate with Heinrich events. This evidence has helped prompt current research on the global ocean circulation system (the halo-saline 'Conveyor', or Global Conveyor), which may be the mechanism for Northern Hemisphere, and perhaps Southern Hemisphere, climate shifts. What causes changes to the 'conveyor' is uncertain; it might be internal or external to the Earth. The process is thought to be that there is a flip-flop fluctuation (with sudden changes) between relatively stable conditions when cold water descends in the North Atlantic, and a situation where this is switched off (the resulting weakening of the Gulf Stream would chill Europe and North America (Broeker *et al.*, 1999).

There is a real possibility that there are quite sudden and marked climatic changes roughly every 1500 years. Current anthropogenic warming might bring the next natural shift forward. If the shift causes a sudden diversion or 'switching off' of the Gulf Stream, Europe and eastern North America would suffer at least 5°C falls in winter temperatures. Some palaeoecologists are convinced that such a Gulf Stream loss happened during the Younger Dryas cold phase (*c.* 12,800–11,400 BP); dendrochronology studies of Irish oak stumps recovered from peat bogs suggest dates for other sudden warm–cold shifts. The threat is so serious that it deserves research and possibly contingency planning.

Most climate change predictions are that the world is warming and sea-levels will rise in coming decades. However, a few dissenters warn that warmer conditions could mean more snow in Antarctica and the Arctic, leading to accretion of the ice caps, which may in turn reflect more sunlight, leading to cooling and falling sea-levels. Warming of the seas could cause increased cloud cover; if this was at high altitude it might chill the world, if at lower altitude it could accelerate global warming. Even as little as a 2°C increase in global mean temperatures would probably mean severe droughts for the tropics and sub-tropics, a 4°C rise would extend droughts to mid-latitudes. Seas warmed above 26°C may spawn more cyclonic storms, and would probably upset present ENSO and NAO patterns. Sudden natural climate changes have happened in the past; anthropogenic warming is no guarantee that such shifts will not recur – indeed, it might make them more likely (Adams *et al.*, 1999). (Goudie, 2002: 174–5 reviewed the causes of sudden climatic change).

5.11 Altered solar radiation receipts

Human influences are now superimposed upon natural stratospheric ozone changes. That was first proven after 1987 when the Antarctic 'hole' became apparent. A number of natural processes can deplete ozone, and the threat posed is made worse because of the reductions human impacts have already caused.

The sun has been relatively stable during recorded history although, every 11 years or so, radiation increases during periods of greater sunspot activity and occasionally there are short-lived bursts of intense radiation associated with large solar flares. Increased receipt of solar radiation may manifest as an electromagnetic storm on Earth. Whether much larger solar flares or periods of much greater sunspot activity can take place is not known, but there may have been such events in the past and they might happen again. There have been speculations about sunspot activity and climate conditions (see the section on the Little Ice Age in Chapter 3).

In 1989 there was sufficient solar activity during a 'normal' solar storm to damage the electricity grid in Quebec (Canada), and there was also disruption of orbiting communications and military satellites. There have been unproven reports that severe influenza epidemics correlate with solar activity; increased radiation might reduce human immunity or raise the rates at which viruses mutate.

Even if the Sun's radiation output remains constant there could be a breakdown in the Earth's shielding caused by the Earth's varying magnetic field, extraterrestrial body impacts, volcanism that injects chlorine compounds into the upper atmosphere, and possibly ocean plankton generating ozone-scavenging compounds. Solar radiation receipts may also be altered by dust generated by human activity, the retreat of ice sheets, or bushfires and droughts – if it settles on ice caps and warms them by reducing their reflectivity. There are signs of such 'dirty ice' in the Arctic at present, mainly as a consequence of industrial pollution.

5.12 Variation in non-solar radiation striking the Earth

Increased radiation on Earth may result from the explosion of a 'nearby' star: a supernova. This could irradiate living organisms directly through the gamma-ray burst or upset the Earth's magnetic field or ozone layer and allow radiation from our Sun to penetrate the atmosphere better. Although supernovae are rare events over geological timespans there is a chance that they may occur close enough to pose a real threat.

5.13 Close encounters of a nasty kind – extraterrestrial body strikes

Most societies fear comets and similar phenomena, either as direct threats or, more likely, as ill omens. There has been speculation that some myths might be based on actual disasters before historical records were kept. In the West the idea that solid heavenly bodies might actually strike the Earth only dates back to William Whiston in AD 1696. In 1796 Edward Topham gave a detailed report of a meteor strike and the recovery of the still warm body from a Yorkshire hillside (UK) to *The Times* (London) (a meteor is the visible passage of a body through the Earth's atmosphere; a meteorite is the 'landed' meteor). That rocks could fall from space was greeted with some scepticism. Plenty of recently landed meteorites have since been collected (some several kilograms in size). Meteorites derived from Mars, the Moon and probably other bodies, have been recovered from ice caps and deserts around the world. After the English Astronomer Royal, Edmund Halley, discussed the possibility of catastrophic strikes in the seventeenth century (Napier and Clure, 1979), interest seems to have increased. In 1822 the English poet Byron apparently suggested that humans might need to destroy comets to survive, and since the start of the twentieth century bodies like the British Interplanetary Society have prompted further interest.

On 30 June 1908 there was a huge explosion at Tunguska (Siberia), an estimated 10- to 40-megaton blast 5 to 10 km above ground. Scientific expeditions were dispatched and they interviewed terrified nomadic peoples and discovered that over 3000 km² of forest had been flattened in a strange pattern reminiscent of damage seen when nuclear weapons have been detonated. The event was probably an airburst caused by a comet (perhaps mainly ice). Another quite large impact took place in Brazil in 1931, and there appear to have been similar blasts in Siberia in 1947 and Greenland in 1997.

Science fiction films and novels about impacts began to appear in the 1940s, 1960s and again in the 1990s. The latter were possibly prompted by events in 1997, when Comet Hale-Bopp was, in most parts of the world, clearly visible to the naked eye for months. With a *c.* 40 km diameter 'solid' nucleus and a 'tail' around 110 million km long this was quite a large and visually impressive 'near-Earth object' (NEO). Comet Hale-Bopp came no closer than roughly 250 million kilometres (it will approach again on an altered orbit in about 2300 years). In July 1994 Comet P/Shoemaker-Levy 9 split into over 21 fragments while approaching Jupiter; many were over 1 km in diameter and caused huge disturbances in the Jovian atmosphere (Jupiter is about 300 times as big as Earth).

Luckily for life on Earth, the larger planets of the Solar System capture many of the bodies that could pose a hazard – although some still get through. Strikes may have been important, helping to 'stir the evolutionary pot', possibly delivering water to Earth, and even precursors of life forms, according to the panspermia hypothesis proposed by Svante Arrhenius in 1903 and promoted by a number of authorities in recent years, including Hoyle and Wickramasinghe (Wickramasinghe, 2001). By causing extinctions, impacts may punctuate stasis, creating new niches and opportunities for survivors to occupy and, through raised UV radiation levels, helping cause mutations that assist evolution. If life is found elsewhere in the Solar System or beyond, one of the things biologists will have to clarify is whether there have been separate origins or whether organisms were carried from solar system to solar system and 'splashed' from planet to planet. Recent studies suggest that the transfer of viable bacteria across even interstellar space might be possible. Some terrestrial micro-organisms might be capable of withstanding vacuum, the cold, and harsh radiation of space, and the accelerations, heating and impacts encountered during re-entry, if they hitch a ride in a large chunk of ice or deep inside a stony meteorite.

Bodies in the Solar System have repeatedly been struck by planetesimals – comets, asteroids, meteorites, bolides and smaller particles – especially before about 390 million years ago when bombardment seems to have been more intense. Astronomers have seen strikes on the moon: in England, Gervase of Canterbury recorded what may have been one in AD 1178; and in 2000 another impact was observed. A small percentage of impacts on Earth have survived weathering and show as craters (astroblemes); one, Lake Manicouagan (Quebec, Canada, *c.* 215 million years BP), is over

151 km in diameter. Other large craters include the Barringer crater (impact around 49,000 BP, Arizona, USA). In 1994 spaceborne imaging radar on the Space Shuttle *Endeavour* found a chain of impact craters in Chad, some more than 15 km wide; they have been dated to around 350 million years BP. The Manicouagan crater is of similar age to at least four other large craters in France, the Ukraine and the USA. If these are plotted on the likely configuration of Asia and North America in the late Triassic when the strikes occurred they form a chain. So, more or less simultaneous multiple strikes (when a body splits up on approach) or a barrage, each strike days or even years apart, are possible (Spray *et al.*, 1998). A cluster of bodies striking over a relatively 'short' span of time in different regions – during anything from a few to thousands of years apart – might explain some gradual, stepped or 'flickering', rather than sudden, mass extinctions.

Many geologists believe that strikes on Earth that have left no visible craters can be traced from fall-out debris thrown up by the impacts. These contain 'shocked mineral grains', aerodynamically moulded pellets – distinctive impact droplets (microspherules, of usually less than 1.5 mm diameter) and larger tektites – soot and charcoal particles, and sometimes iridium particles (a rare metal that might have come from extraterrestrial sources, rather than volcanoes). The KT-boundary is marked by a widespread thin clay layer, indicating abrupt change in conditions; the base of the layer is rich in iridium (Alvarez *et al.*, 1980; 1982; 1984). The hypothesis is that an extraterrestrial body struck (probably near present-day Yucatan, Central America) and threw up iridium amongst the dust and debris, which upset the environment enough to cause a mass extinction. This has been challenged by those who suggest that widespread 'global' wildfires were the cause of environmental damage and extinctions (but this does not explain the iridium). Wildfires on a number of continents could be a consequence of an extraterrestrial strike but were common during the Late Cretaceous, and were probably started by earthly causes (Scott *et al.*, 2000). A large impact would raise huge quantities of debris and aerosols into the Earth's atmosphere, whether it took place on land or at sea, causing acid deposition and lowering global temperatures sufficiently for long enough to catastrophically affect living organisms (Toon *et al.*, 1982). Iridium anomalies have been found in marine and terrestrial KT-boundary sediments, which seems to rule out the possibility that marine algae could be able to concentrate the metal; but there have been accounts of high levels of iridium associated with an eruption of the Kiluea Volcano (Hawaii) in 1983. So it may be possible that volcanic outpourings, fed by much deeper magma reservoirs than most of today's volcanoes, spread iridium through the atmosphere at the close of the Cretaceous. Perhaps both impacts and volcanoes have caused mass extinctions.

The Earth is constantly under a barrage of smaller than sand grain-sized extraterrestrial objects, and larger strikes are by no means uncommon. McGuire (1999: 208) estimated the frequencies shown in Table 5.1).

Table 5.1 Estimated strike rate of extraterrestrial objects

Object size (diameter)	Frequency of impact with Earth
Pea	5 minutes
Walnut	60 minutes
Soccer ball	1 month
50 m	50 to 100 years
500 m	10,000 years
1 km	100,000 years
2 km	500,000 years
10 km	50 to 100 million years

In early 2001, NASA mapped about 150 impact craters on Earth, almost all less than 600 million years old, and the bulk between about 2 km and 20 km in size; smaller craters are less easy to identify and many more are likely to be found as research continues (for a map see http://observ.ivv.nasa.gov/nasa/exhibits/craters/impact_earth.shtml.html, accessed 27 March 2001). Shoemaker (1998) suggested that impacts with the Earth may have *increased* in the last 100 million years, possibly reflecting the passage of the Sun through the Galactic plane over the last 65 million years.

Debris continually collides with the Earth, with particles bigger than a sand grain visible as meteors, and smaller particles settling slowly as dust. Anything over about 10 metres in diameter is likely to cause serious local surface damage (Steel, 1995). An asteroid or comet of 50 metres in diameter could destroy a city. Something of 100 metres in diameter would cause damage at a regional scale, the seriousness depending on its composition, speed, location of strike and angle of approach. A body of 1 km diameter would probably kill several hundred million people, no matter where it impacted; such a strike is statistically likely every 100,000 years. Bodies sometimes explode at height; the blast at Tunguska from quite a small body was probably equivalent to a large atomic weapon.

Evidence from ancient texts, supported by dendrochronology, hint at an episode of asteroid or comet strikes around 2350 BC large enough to damage civilizations around the Mediterranean, in the Middle East and China. British 'Arthurian' legends and the Anglo-Saxon epic poem *Beowulf* may have been influenced by a bombardment between AD 530 and 540, and tree ring evidence suggests cold conditions between AD 536 and 540. Several British chroniclers mention 'dry fogs' around AD 536–537, so something seems to have happened in the mid-sixth century, whether it was a strike or volcanic is not clear. Maori histories tell of large impacts before the arrival of Europeans (in the sixteenth century), and there are Tunguska-like patterns of tree felling near impact craters at Tapanui (South Island, New Zealand); some environmental historians have gone so far as to blame

Polynesian migrations on the disruption caused by these events (ENSO has also been blamed).

Science has taken more interest in extraterrestrial body strikes since papers were published by Walter Alvarez and others on the KT-boundary mass extinction. Alvarez suggested the cause was a body (an asteroid or comet) of about 10 kilometres in diameter; its possible impact site has been identified on the Yucatan Peninsula, Mexico: the Chicxulub 'crater'. There is also evidence of similar-age tsunami debris far north into the USA, as well as global fall-out, including the distinctive iridium-rich layer.

A rocky or metallic asteroid as small as *c*. 1 kilometre in diameter is capable of causing a serious catastrophe by blast, earthquake, tsunami, a near-global shower of hot fall-out that would ignite bushfires, a global dust-veil effect, probably ozone layer disruption and possibly volcanic activity triggered by the shock. Bodies of this size may well strike every 100,000 years or so. In the late 1990s the Sandia National Laboratories (USA), a body well versed in the ways of nuclear weapons blasts, computer-simulated an ocean strike by a 1.4 km-diameter rocky asteroid. The result was a near global catastrophe (see http://www.sandia.gov/media/comethit.htm, accessed 30 April 2001). A larger asteroid of 5 to 6 kilometres in diameter impacting in the mid-Atlantic would probably generate a tsunami reaching several hundred metres in height onshore; the USA would be badly damaged as far as the Appalachians, as would much of Europe west of the Alps and Pyrenees (Hills and Goda, 1999); in addition, there would be other impacts like blast and dust-veil cooling.

In 1937 the asteroid Hermes passed close to Earth (at 500,000 tonnes, a strike would have been severe); another 'near-miss' by a *c*. 50-metre asteroid took place on 23 December 2000; in January 2002 a *c*. 300 m-diameter asteroid 'near-missed' the Earth (and was not spotted on the approach), and a 'close-pass' by another of 30 to 70 metres in diameter is likely in 2030. In AD 2126 comet Swift-Tuttle will pass close-by. Since the late 1970s several bodies of 5 to 20 metres in diameter have 'near-missed' Earth and, in 1989, a 300 m-diameter body (Asclepius 198FC) came within a distance equivalent to that at which the Moon orbits. So, it is a little surprising that until recently the threat was seen by most to be more science fiction than real. However, there is a great deal of evidence of past strikes and near-misses. ('Near-miss' effectively means a roughly 1 in 500 chance of a hit, quite a high risk in comparison with many accepted 'real' threats.)

Recent estimates vary somewhat, but it seems reasonable to put the odds of a body of 1 to 10 kilometres in diameter striking the Earth at roughly 1 per cent over the next 1000 years (Rabinowitz *et al*., 2000); the probability of a body of up to 60 metres in diameter striking is around 1 in 20,000. These risks may be compared with the chances of an average person in the UK suffering death by motor accident (1 in 1000), by aircraft crash (1 in 20,000), by flood (1 in 30,000) and by tornado (1 in 60,000) (Muir, 2001). One problem is that something passing through the Solar System past Earth, even if it misses, is likely to be on a return orbit, and that orbit will have been altered by the pass, making future predictions difficult.

What can be done about the risk from extraterrestrial bodies?

A recent review of websites disclosed a number of worried people urging monitoring and early warning, plus planning for mitigation or avoidance (see Box 5.1). A letter to *New Scientist* (7 April 2001: 54) noted that, even though few – if any – have been directly killed by H-bombs, the threat of these is taken very seriously and absorbs vast sums of money. Even a smallish body of 1 to 2 kilometres in diameter could easily destroy half the human population of the world. A grading system of risk has been developed: the Torino Impact Hazard Scale of 0–10. The scale is derived from size and likelihood of a hit (Binzel, 2000). A score of 8–10 is given to an object pretty certain to impact with Earth and big enough to cause a serious threat.

In addition to direct strikes, impacts with the Moon or some other planet might have a 'billiard ball' effect – say, ejecting large quantities of small-sized debris or larger fragments that could then strike Earth.

Box 5.1 Possible ways of reducing danger from extraterrestrial bodies

Avoidance
- Monitor for NEOs while they are at as great a distance from Earth as possible, to give adequate time for response (this is only likely to be partly effective because some objects approach fast and from 'out of the sun').
- In addition to costly programmes by national and international bodies, amateurs can be used to aid monitoring, via the Internet, for a low outlay. Another possibility is for military surveillance units to spend some of their time monitoring and training on NEOs/planetesimals.
- Map and track known objects.
- Develop devices to nudge away, deflect or destroy threatening objects.
- There have been suggestions that the proposed USA missile defence shield could also be made to protect the Earth against NEO strikes.

Mitigation
- Store food reserves (a wise precaution, in any case, against crop failures and other natural disasters, or terrorism and unrest).
- Duplicate and disperse critical collections of genetic resources, important records and library resources (ideally at high-altitude mountain sites to reduce tsunami risk).
- Harden crucial storage sites for dangerous and important materials; duplicate the latter (e.g. collections of seeds, libraries, data banks).
- Prepare contingency plans for evacuations, distribution of emergency supplies, civil defence.

In 1990 the US House of Representatives charged NASA to investigate the threat of strikes, the improvement of detection (early warning), and to explore ways of altering NEO approach trajectories or destroying them. NASA reported back in 1993, calling for several large optical telescopes to be dedicated to searching. Since 1990 a number of bodies – including CSIRO (the Commonwealth Scientific and Industrial Research Organization, Australia), the Russian Parliament and the British Parliament – have discussed ways of responding to the threat. While enthusiasm for avoidance has been growing, the efforts are underfunded and it is far from certain that the public in the West, let alone in poorer countries, are really convinced of the need for expenditure. By 1994 the USA had progressed via working groups and studies to provide some (limited) funds for risk assessment: the Spaceguard Survey. In 1991 the International Astronomical Union set up a working group to investigate the threats; four years later at a meeting in Vulcano (Italy) the group decided to establish an organization to coordinate and sustain research efforts worldwide: the Spaceguard Foundation. Unfortunately progress is limited by lack of money (Tate, 2000). The UK decided (in August 2001) to establish a research centre to study the risk of NEOs/extraterrestrial body strikes, based at Leicester University. Unfortunately, the Near Earth Object Information Centre seems likely to be more of a public relations exercise than an adequately funded effort (see www.nearearthobjects.co.uk, accessed November 2001).

Gribbin and Gribbin (1998: 3) calculated the annual risk of an impact big enough to cause global crop failure (something that could probably happen roughly every 300,000 years) as 1 in 1.2 million; assuming a human lifetime to be 75 years and the present human population 6500 million, the risk to an individual in any year is 1 in 2 million. That would mean, *if calculated as an average*, 2700 deaths per year (390 of them in developed countries). The cost of road safety measures in the UK in the mid-1990s was around £800,000 (*c.* US$1,240,000) for each life saved – so, using similar costings, to prevent at least equivalent threat from NEOs, the developed countries ought to be spending at least US$500 million a year. Nowhere remotely near that sum is being allocated. Table 5.2 (based on a list presented by McGuire, 1999: 40) gives some indication of the risk of death by asteroid or comet strike, compared with various 'common' causes (faced by an average American).

If an incoming NEO is detected on a collision course it may be possible to deflect or destroy it using technology already available. The problem is that efforts to destroy a body larger than about 1 kilometre in diameter might split it into several dangerous fragments. There are likely to be very different effects caused by exploding atomic bombs near an NEO, depending on whether a body is solid rock, split by fissures, loosely joined rubble or ice. More knowledge is needed to help prepare for deflecting or destroying NEOs. In 1997 NASA sent a probe (the NEAR Shoemaker craft)

Table 5.2 Risk of death by asteroid strike, compared with various 'common' causes

Cause of death	Risk
Car accident	1 in 100
Murder	1 in 300
Fire	1 in 800
Electrocution	1 in 5000
Asteroid/comet impact	1 in 10,000
Aircraft accident	1 in 20,000
Hurricane	1 in 25,000
Flood	1 in 30,000
Tornado	1 in 50,000
Earthquake	1 in 200,000

past a 60 km-diameter asteroid called Mathilde; in early 2000 the same craft made a close-proximity survey of the 35 km-diameter asteroid Eros. After circling Eros for a year the craft made an opportunistic 'gentle' crash landing in 2001. Eros and Mathilde seem very different, the latter asteroid may be a rubble body, which would be difficult to deflect or destroy, while Eros seems solid and possibly easier to deal with. In 1999 Clementine-2 was launched to probe asteroids with a view to establishing how they might best be destroyed or steered. Other probes have made close-passes of comets, and NASA has plans to sample a comet nucleus within a few years; the agency also hopes to hit an asteroid with a fast-moving heavy copper 'bullet' (the Temple 1 craft) in 2004, to gather information that will help future efforts to deflect or destroy threats (see http//:deepimpact.jpl.nassa.gov, accessed July 2001).

In 2001 the NASA probe, Deep Space 1, obtained detailed photographs of a comet. Flying within *c.* 2800 km (1400 miles) of Comet Borrelly it revealed a nucleus roughly 10 km long, which appeared too dark to simply be composed of ice (*The Times*, UK, 17 December 2001: 78).

The Spacewatch Project (an ongoing survey started in 1989 using an optical telescope on Kitts Peak near Tucson, USA) will provide some warning, and a 1.8 m-diameter telescope is being constructed to replace the smaller instrument currently conducting monitoring. Radar is probably of little use for early warning due to the distances involved, the speed of the bodies (which would cause doppler shifts), and because background 'noise' hides smaller bodies. Earth-based optical telescopes may miss objects if these approach 'from out of the Sun', or are dark and non-reflecting; the solution might be to place optical telescopes in a similar orbit to Jupiter to get a better view, which would be costly and technically challenging. For comets there may be some hope of better detection methods, but for 'stealth' asteroids, adequate early warning will be difficult to achieve. The risk from NEOs will remain and a wise precaution is to store adequate food and other emergency supplies, and to disperse vital facilities in widely separated, secure (ideally high-altitude) sites.

5.14 Reduced sunlight penetration of the atmosphere

There are a number of things that could cut the amount of sunlight reaching the Earth's surface, causing reduced photosynthesis and cooling, perhaps for weeks or for much longer; humans have added 'nuclear winter' to the list (the effects of dust and smoke caused by large-scale use of nuclear weapons). The natural causes include large-scale forest fires, dust and aerosols injected into the upper atmosphere by volcanoes or extraterrestrial body impacts, and increased cloud cover. Extraterrestrial body impacts and mega-eruptions could directly cause a dust veil and also trigger fires, generating smoke and soot. Following a period of such disruptions there might additionally be stratospheric ozone losses leading to increased ultra-violet (UV) irradiation. Organisms adapted to undertake underground hibernation, deepwater and cold terrestrial conditions would stand a better chance of survival.

The present human population would find it difficult to survive serious 'global winter' conditions for more than a few months, even if food reserves were maintained at higher levels than present. Prepared survivalist groups and small isolated communities like oceanic islands with access to good food reserves and possibly deep-sea fishing might survive.

5.15 Diseases that pose a threat

Crop and livestock diseases

Crop and livestock diseases have seriously affected humans, primarily by causing hunger. These diseases also disrupt rural economies – as did the UK's 2001/02 foot and mouth disease cattle epidemic – and can determine what settlement patterns and livelihood strategies are viable. Livestock may be a source of infections that 'jump' species and infect humans (notably influenza, various encephalitis strains and pox viruses). In 1889 Italian troops occupying Somalia accidentally introduced the animal disease rinderpest to Africa; within a few years it had spread to decimate cattle and many wildlife species across Africa (see also the discussion of tsetse flies later in this chapter).

Human diseases

A wide range of factors affect disease transmission; these include psychological, behavioural, environmental, dietary conditions, exposure to disease vectors, suppression of immunity and misuse of antibiotics. Diseases are often triggered or hindered by environmental factors, but they are also much influenced by human activities, so it can be difficult to separate physical and anthropogenic causes and controls. Since the early 1990s there has been growing interest in predicting the health impacts of climate change. Much of this has been prompted by worries about the effects of global (anthropogenic) warming (McMichael *et al.*, 1996). Rising population reflects the fact that humans have managed to improve their food supplies and control many diseases. The development of vaccines and antibiotics

Figure 5.4 A mega-city, prone to earthquake, and with nearby volcanic activity: central Mexico City
Source: Author, 1998.

since the 1900s has had particular impact. However, new diseases can appear (at least 30 have done so since 1976) and known infections can alter their virulence and patterns of transmission; also, today, far more people travel and at much greater speed than ever before. Another problem is the expansion of cities, which may outstrip infrastructure and service provision, and results in overcrowding and often impoverished groups living in substandard, insanitary housing. The transmission of disease is still a threat.

Diamond (1997) adopted a deterministic argument, that biogeographical factors – notably disease transmission and immunity – helped Europeans dominate the modern world ('McNeill's Law'). Similar arguments have been aired by Crosby (1986) and others. Historically, four human diseases have probably had greatest impact: the Black Death (generally assumed to have been bubonic plague), influenza ('flu'), smallpox and malaria. Cartwright and Biddis (2001) caution that it is not just the killing of large numbers of people that affects human fortunes, there is also the trauma of an epidemic that changes attitudes, decisions made by diseased or unstable administrators, and ongoing labour shortages and cultural changes, which often hit food production and lead to debilitation of survivors.

BUBONIC PLAGUE

'Plague' has been a catch-all term for epidemics; sometimes illness has been more specifically attributed to Black Death. This is widely assumed to have been bubonic plague (*Yersinia pestis*). Until comparatively recently, plague has been a significant control on human numbers; there have been three recorded worldwide epidemics after which the disease reappeared from time to time. Serious outbreaks occurred in Mediterranean countries in the third century and possible earlier, and in the sixth century, during the reign of the Byzantine Roman Emperor Justinian (around AD 542). Outbreaks in Europe between 1347 and 1350 probably killed about one-third of the total population (roughly 20 million). Some of the 'lesser' recurrences were catastrophic: between AD 1200 and 1400 roughly half the Chinese population was killed (some by famine); and, in 1665 at least 59,000 died in London. In China, in the late nineteenth century a severe outbreak was probably bubonic plague.

After the eighteenth century, improved sanitation, better buildings, quarantine and evacuation of infected areas helped control the disease. During the twentieth century antibiotics and vaccines became available to treat it (assuming it was bubonic plague), plus insecticides to counter fleas, and rodent control to reduce the rat population. Plague is still present: in the 1990s bubonic plague struck in Africa and India, and in some cases was found to be antibiotic-resistant. In 1994 the Indian city of Sarat had a bubonic plague outbreak, which was controlled by prompt use of antibiotics; the cause seems to have been poor refuse collection and sewage disposal, which let to an abundance of rodents. These problems are currently common in many cities in developing and developed countries where too little is being spent on public hygiene.

The reaction of people to past plague outbreaks has been quite well recorded (especially for England), offering modern governments valuable insights into how citizens react to epidemics for which there is little proven treatment. Outbreaks in the fourteenth century are generally thought to have started on the fringes of the Gobi Desert (where there are rodents that harbour the bubonic plague bacterium). In the 1320s disease spread to Europe, India and Africa, and around 1356 reached the Mediterranean, probably among troops and along trade routes (in 1347 Tartars besieging the Crimean city of Tana, now Feodisia, reputedly catapulted victims of bubonic plague over the city walls, one of the earliest acts of biological warfare). During 1347 plague reached Geneva, by 1348 Bristol, and by 1349 all but the remoter settlements of Europe, Scandinavia, Iceland and the Mediterranean countries were infected and probably huge swathes of Africa and Asia.

However, there is still considerable uncertainty about the cause of the Black Death, which repeatedly struck the Old World (the First Book of Samuel in the Old Testament mentions an outbreak that might have been bubonic plague, perhaps 4000 BP). Some of the characteristics of the epidemics recorded do not match what is known about bubonic plague – in particular the apparent rapidity of transmission, the supposed absence of brown rats in some afflicted areas (in the fourteenth century Iceland is supposed to have had no suitable rat vectors), the absence of reports of the mass death of rats, and the puzzling apparent failure of the disease during one major epidemic to spread to Poland (Ziegler, 1990). An analysis of four-teenth-century English church records suggests too rapid a spread and too high a mortality rate for it to have been bubonic plague; haemorrhagic fever of some kind might have been to blame (*The Times*, UK, 13 April 2002: 12).

Sceptics question whether recorded Black Death symptoms match bubonic plague: outbreaks happened in winter and summer, and in urban and rural areas (so in populous and sparsely settled situations); the onset seems to have been too rapid in some cases – might it have been anthrax caught from cattle? Mackenzie (2001) suggested that at least some of the supposed bubonic plague outbreaks might have been a form of viral haem-orrhagic fever like Ebola (a disease still present in parts of Africa). If these fears are correct, the reappearance of such a plague in the modern world is a very real threat, and would probably have worse consequences than it did in medieval times (Scott and Duncan, 2001). So there has been interest in establishing what past plagues were. In 1999 researchers isolated DNA characteristic of bubonic plague (*Y. pestis*, which is very similar to *Y. pseudotuberculosis*) from the dental pulp of plague victims buried up to 405 years ago in Provence and Marseilles (France), proving that bubonic plague killed them (but *not* establishing that it was the main cause of death – some-thing like Ebola could also have been present). Similar studies with many samples from other plague victim burial pits have failed to isolate *Yersinia*. Until the cause of plagues is known for sure, speculation about environmental links must be cautious. Studies of people descended from those who survived infection with Black Death in the UK village of Eyam in the fourteenth century show the former have a high resistance to HIV/AIDS, possibly because they carry a gene that prevents infective organisms from reaching the lymph system (Cantor, 1997: 227–30).

Yersinia bacterium remains in wild rodent populations around the world (including the Americas), and breaks forth occasionally to infect people. A number of environmental historians have linked Chinese outbreaks to climatic changes – cooling, floods and famine – which, they argue, debilitated humans and favoured rodent breeding and movements that brought them into contact with weakened people. Baillie (in Slack, 1999: 62–79) noted some correlation between adverse climate and plague outbreaks in the 1340s, but cautioned that other factors might be involved. It is tempting to correlate outbreaks like those in China in 1588/89 and 1640/41 with strong El Niño events.

Although it is impossible to be precise, the Black Death probably killed 43 million people between 1347 and 1352; some estimates are that close to 50 per cent of the world population at the time died. Whatever the cause, the trauma suffered by the survivors was immense and, together with labour shortages, was a powerful force that put an end to the outlook and many of the ways of the medieval period (Gottfried, 1983; Ziegler, 1990). The European population did not recover to the numbers there had been in medieval times until around 1500. Plague probably helped end feudalism and may have set in train a shift from cheap labour to technology, which supported the Industrial Revolution after 1752 (see Box 3.2). Some go further and argue that plague helped trigger the religious Reformation in the UK, the persecution of European Jewish populations and even the demand for African slaves. From the 1360s until well into the sixteenth century Europe continued to receive visitations of plague, on average every decade or so, and was not rid of it until the nineteenth century (East Asia had a major **pandemic** in the 1890s and there were outbreaks in South Asia in the late 1990s). Toynbee (1934) saw plague as one of the major environmental challenges to European civilization; the LIA must also have challenged it.

SMALLPOX AND SIMILAR DISEASES

Smallpox (*Vaccinia*) is an extremely contagious infection caused by the *Orthopoxvirus variola* virus. It afflicted humans in the Old World throughout history until a few decades ago, causing blindness, disfigurement and death, often of a high proportion of those infected. It probably killed around 300 million worldwide in the early twentieth century alone. Outbreaks were common, and when isolated people came into contact with the infected, severe outbreaks ensued, like that in Siberia in 1630.

Controlled largely by immunization, the last documented case occurred in 1977. Eradication was declared in 1979, but there is a possibility it might reappear from cattle pox mutation, laboratory escape or misuse of stored viruses. Along with anthrax, smallpox is considered one of the most threatening bioterrorism risks (Rosen, 2000; Ligon, 2001).

INFLUENZA

There have been three great influenza (flu) pandemics in recent history: AD 1918, 1957 and 1968. In 1918 Spanish flu (the 'Spanish Lady' outbreak) spread worldwide, infecting most of the human population, and by 1920 had killed over 40 million – roughly four times the number killed in battle during the First World War. The victims often included healthy, fit and young people. The 1918 pandemic finished quite quickly and the flu strain seems to have effectively disappeared as people developed immunity against it. The movements of troops between countries and within Europe, relocation of refugees, poor sanitation during wartime and the inadequate diet of both military personnel and civilians probably aided its spread (Aronson, 2000). Less serious pandemics occurred in 1957 (Asian flu) and 1968 (Hong Kong flu); nevertheless, the former killed about 70,000 and the latter 36,000 worldwide. Most epidemiologists feel that serious flu pandemics (especially the A-strain) are a very strong possibility in the future.

It seems that the virus constantly re-sorts its genes, sometimes 'adding' new material from animals, especially goats, pigs, mice, poultry and cage birds, geese, gulls and shore birds. The chances of new virus strains appearing are enhanced where people and suitable animals are brought into contact during farming, marketing or by living in close proximity. Birds seem to be the main carrier of the more dangerous avian (A-type) flu viruses, and direct infection or infection via pigs may be possible. Whether changes in flu virulence are increased by environmental factors like floods or unusual weather conditions is not clear; these may certainly cause movements of people and wildlife, and debilitating famine.

In Hong Kong in 1997 a strain of avian flu spread to humans from poultry, probably having originated in southern China; it killed six people in Hong Kong (a mortality rate of about one-third). The 1997 Hong Kong virus had the potential to be as virulent as the 1918 strain; luckily the virus has so far failed to develop effective human–human infection, and the Hong Kong authorities immediately slaughtered all potentially infected chickens in the territory. Recently there was concern that pet parakeets in India may pass on a dangerous flu strain; certainly, it would be wise to tighten controls on the trade in cage birds (Agrawal, 2001), and monitor chicken and other domestic fowl production.

It seems likely that modern populations would be little more prepared for a new and virulent flu strain than those of 1918: the sheer number of victims would probably overwhelm health services unless an effective vaccine could be prepared and administered in advance or very soon after an outbreak. Because A-type flu mutates quickly, vaccines are difficult to prepare and administer fast enough. With modern transport, the spread will be quicker than it was in 1918 or 1957, making vaccine production and distribution a challenge. Scientists have been exhuming victims of the 1918 pandemic buried in permafrost conditions, to try and sequence the DNA of that virus and prepare for similar strains in the future; so far, with limited success.

MALARIA

There are two forms of malaria, caused by different protozoans: *Plasmodium falciparium* (clinically the more dangerous form) and *Plasmodium vivax* (less dangerous but more widespread). Malaria is transmitted by many different mosquitoes, although the main vectors are the *Aedes*, *Culex* and *Anopheles* species. These mosquitoes vary in their habits and preferred conditions, consequently environmental changes may favour malaria transmission by one and not another. As for other insect-transmitted diseases, there are a number of factors affecting transmission: the insect involved, the survival or propagation of the disease organism inside the insect; environmental factors that alter human behaviour, and thus exposure to insects; cultural factors that alter exposure (through changed housing conditions); environmental changes that bring insects into contact with infected reservoir organisms (not so important for malaria, but markedly so for yellow fever and a range of other diseases); and pollution, which may discourage insect breeding or facilitate it (as in the case of dengue where litter provides some of the mosquito breeding sites).

Fevers, probably malaria, have been recorded throughout history in the Old World, and it was probably introduced to the Americas in the last 500 years. There is archaeological evidence that, towards the close of the Roman Empire, farming areas may have been devastated by falciparium malaria outbreaks. The greatest changes in infection rates took place as building standards improved and following the introduction of quinine to Europe in the 1640s, and to India and Africa after the late 1650s. More widespread use of quinine began in the 1850s, and improved anti-malarial drugs, especially the chloroquines, helped reduce transmission after the 1940s. Quinine was used to treat the symptoms of malaria long before it was known how the disease was transmitted; it was not until 1897 that Ronald Ross discovered mosquitoes were the carriers of malaria. Eradication was achieved in many regions by the late 1950s. Unfortunately, malaria has since become increasingly resistant to chloroquine drugs and mosquitoes to insecticide; added to this, poor health planning, shortage of funding and civil unrest set back malaria control. In 1972 the Global Eradication of Malaria Program was abandoned by the World Health Organization (WHO). Honigsbaum (2001) estimated that the global death toll from malaria over the period 1986 to 2001 was around 28 million, and roughly 500 million catch the disease every year. Some experts fear that the disease is now a growing threat. An anti-malarial drug (*qinghaosu*) derived from a herbal fever treatment long used in China may prove effective against chloroquine-resistant malaria, and there are now hopes of new vaccines against the malaria parasite.

Linking environmental causation to malaria incidence is difficult given the many factors affecting transmission, especially the many different species of mosquitoes involved. There have been correlations between El Niño events and outbreaks (like that in the Punjab, India, in 1908), suggesting that these favour mosquito breeding and transmission; proving these suspicions and moving to accurate forecasting is less straightforward. Considerable efforts have focused on predicting future patterns of malaria transmission, assuming that global warming or shifts to wetter conditions will favour transmission; for example, spreading to uplands in the tropics and to higher latitudes (see Parry and Carter, 1998: 44). However, while climatic changes can alter the vector organisms' distribution and behaviour, and the incubation period for the parasite within them, it also affects human habits, housing and clothing, which changes exposure. If more pools of water form, mosquito breeding may increase but, equally, more rain can sometimes flush away mosquito larvae.

Anopheles mosquitoes are effective at transmitting malaria when temperatures are between 20° and 30°C, but can breed when it is as cold as 16°C (the minimum temperature threshold for *vivax* malaria transmission is several degrees lower than for *falciparium*). The marshlands of England, France, The Netherlands, Germany and many other parts of Europe had endemic malaria in the past. Global warming, whether natural or anthropogenic might pose the threat of a reappearance of malaria and outbreaks in new areas. However, as discussed in Chapter 3, the UK's infection patterns during the Little Ice Age may not support some of the assumptions. There are also different possible transmission scenarios: at one extreme, year-round transmission occurs and the exposed population develop a relatively high immunity, so that deaths mainly strike among children, the debilitated and pregnant; at the other extreme, transmission is infrequent, immunity low and, from time to time, there are outbreaks with high mortality for all segments of the population.

Modern transport can spread virulent strains of malaria (through infected passengers or 'hitchhiking' mosquitoes) to new areas where people have little resistance. In the past ship travel was usually slow enough to reduce the chance of spread. In 1930 *Anopheles gambiae* mosquitoes reached Brazil from West Africa by aircraft; 20,000 or more people died between 1938 and 1939 before the outbreak was controlled. Compared with the 1930s the potential for spreading malaria by air has greatly increased (for further information see www.malaria.org, accessed November 2001).

YELLOW FEVER

Yellow fever was prevalent in the southern USA until the early twentieth century; roughly 10 per cent of the population of Philadelphia died from it in 1793, and in 1877 the Memphis (TN) region was badly hit. It is still present in South and Central America, where various monkey species act as a reservoir of infection from which it is occasionally spread by *Aedes* spp. mosquitoes, perhaps causing an epidemic if human-to-human transmission is established. In 1995 an outbreak hit Latin America (especially Peru and Amazonia), spread by a mosquito species that can thrive in urban areas. Transmission is very much affected by temperature and humidity (which must exceed 20°C with high humidity), and is often triggered by forest clearance and population movements that bring people close to mosquitoes. The disease originated in Africa, and was probably carried to the New World after the Conquest, and to tropical Asia and Southeast Asia. There are rare cases of transmission in Europe, and non-tropical Asia and Australia where the infected mosquitoes have been transported during warm weather. Any rise in temperature could mean that the disease becomes endemic in new areas. There have been claims that some outbreaks correlate with ENSO events (like the 1877 Memphis epidemic), but this is unproven.

DENGUE

Dengue (haemorrhagic or 'breakbone' fever) is caused by four types of virus related to that which causes yellow fever. Severe illness is associated with recovery from infection by one type of dengue virus and then infection with another. It is transmitted mainly by *Aedes aegypti* mosquitoes (and also by *A. albopictus* in Asia and, recently, the Americas between Chile and Canada), which breed in small pools or water-filled containers, guttering, litter-strewn areas, etc. There is no vaccine available. Dengue is encountered in urban environments worldwide where temperatures remain above 10°C. The disease is especially common in Southeast Asia and, after signs of improvement in the 1970s, there has been a marked expansion recently into new areas. Some dengue outbreaks have been correlated with ENSO events, although this link is not proven (Grove and Chappell, 2000: 97–8). The international trade in used tyres may have helped spread dengue to New Zealand, the Americas, Africa and Europe (the tyres are left standing where the disease is endemic, and carry infected mosquitoes and larvae to new areas).

DIPHTHERIA

Although controlled in most developed countries, there has been a marked increase of the disease in the former USSR as a consequence of governance changes since 1992.

TUBERCULOSIS

Tuberculosis (TB, *Mycobacterium tuberculosis*) is one of the diseases apparently 'tamed' during the twentieth century by antibiotics, immunization and mass screening, at least in the developed countries. The disease may well infect about one-third of the world population. Recently, it has been making a comeback in a virulent drug-resistant form (multi-drug resistant (MDR) strain) in many developing and developed countries. Chemotherapy of TB requires the victim to undergo a complex and costly daily drug regimen. While this may be possible in more affluent countries with cooperative patients, it is difficult to administer to vagrants and drug addicts, and in poor countries there are not the resources available. Current anti-TB medication is costly, may be poorly stored, and is sometimes badly manufactured or counterfeit. Even developed countries now have very poor drug treatment compliance rates: in 1991 New York City had a rate of 10 per cent, lower than all developing countries, yet there was a relatively high rate of infection (as much as 51 per cent in some poor communities). Without great care in any country a fair proportion of those treated will fail to complete their dosage, will pass on the infection, help promote drug-resistant strains of the illness and continue to suffer. The infected may show few symptoms and can spread TB in crowded buildings, and on public transport and aircraft, especially if air conditioning has inadequate filtration. TB probably kills roughly 3 million a year, but significantly contributes to the death of far more, especially those with HIV/AIDS.

In April 2001 the UK had cases of an especially virulent strain, first detected amongst school children in Leicester, which was easily transmitted even to some who had already been vaccinated. Countries should invest in mass screening and thorough community nursing if they are to detect infection, and ensure adequate drug treatment and control of transmission.

MEASLES

Measles was a serious scourge in rich and poor countries until after the Second World War, and is still far from controlled. It is also easy to overlook the impact that common childhood diseases of Europeans have had upon peoples in other parts of the world. In 1875 the son of a Fijian chief carried the disease back to the islands from a visit to Australia and within four months 20,000 had died. The disease had a similar impact in Latin America and North America soon after Europeans arrived.

Nowadays, measles kills about a million children a year worldwide (mainly in poorer countries). Many survivors are left permanently debilitated. Without current levels of vaccination the death rate would be far higher. The disease mutates, making available vaccines less effective, and it is acquiring antibiotic resistance. Once the WHO talked of eradication, now it aims to halve deaths by 2005. Vaccination needs to be expanded fast, both to cut the misery being caused, and to reduce the risk of a pandemic of new virulent and difficult-to-treat strains. However, fears that the widely used measles/mumps/rubella (MMR) vaccine can lead to autism has led parents in some countries to avoid vaccination. It is a disease that could flare up amongst populations weakened by famine or concentrated in refugee camps.

TYPHOID

Typhoid is an acute, highly contagious disease caused by the bacterium *Salmonella typhi*. It is transmitted mainly through poor hygiene, contaminated water and food. Control of flies and other pests, adequate sewage disposal and good hygiene practices are important for the control of the disease. Typhoid often follows natural disasters when flooding leads to drinking water and food contamination, and where people are concentrated under poor conditions such as refugee and squatter settlements. In temperate developed countries the disease is normally kept under control but epidemics could easily develop during natural disasters or civil unrest.

CHOLERA

Cholera has similar transmission patterns to typhoid and is caused by the bacterium *Vibro cholerae*, ingested in water, food and occasionally dust. The bacteria can survive well in the environment, especially in shellfish and copepods, and can reappear after human carriers have been treated or removed from the population – outbreaks are often seasonal. In recent years the disease has spread in areas of urban squalor and it has become more common in Latin America since 1991. The 1990–95 El Niño event is believed by many to have favoured outbreaks in Peru, and to have made its control more difficult and costly. Global climate change may alter the incidence of cholera (Borroto, 1998).

ESCHERICHIA COLI

Occurring globally, the transmission of various strains of *Escherichia coli* (*E. coli*) can be affected by weather conditions, natural disasters and pollution. Warm weather, poor food hygiene and consumer fashions play a major role in outbreaks.

LOUSE-, TICK- AND FLY-BORNE DISEASES

There have been increased rates of transmission of Lyme disease in the USA due to changing land uses and, possibly, rising temperatures. Deer carrying the ticks that spread the *Borellia burgdorferi* bacteria and people now come into closer contact than they did a few decades ago due to housing and farming changes. The likelihood is that global warming will favour the transmission of tick-borne and louse-transmitted diseases. Lyme Disease, like typhus, is caused by species of the *Rickettsia* bacteria (*R. prowazeki*).

Typhus is a lice-borne disease that has often appeared in times of famine and migration; it did so during the famines in Ireland in the 1840s. It spreads where people are concentrated under squalid conditions – in Europe, even during cool parts of the Little Ice Age, the disease frequently appeared when there was warfare that generated refugees. Lice are increasingly pesticide-resistant and are by no means controlled, even in developed countries. Warm conditions may favour typhus a little because then the disease can also be inhaled on dust. There are several *Rickettsia* bacteria that are quite vector-specific and variable in virulence; they have been given regional names, such as Rocky Mountain Spotted fever and Q-fever. The incidence of these diseases is probably in part related to temperature conditions.

A total of 22 species of tsetse fly (*Glossinia* spp.) spread sleeping sickness in tropical Africa. The causative trypanasomes (*Trypanosoma gambiense* and *T. rhodesiense*) infect livestock, some game animals and humans. Sleeping sickness has long been a serious problem, causing illness and debilitation, and making large areas of tropical rangeland no-go areas for ranching. It currently affects about 25,000 people a year in central Africa alone. Tsetse flies are sensitive to temperature and rainfall. The fly needs good vegetation cover, which develops in many regions only in wetter years. Even a slight temperature rise or change in rainfall patterns might spread it to areas that are presently free because their altitude or latitude make them too cold or dry. In 2002 the UN started supporting the Pan-African Tsetse and Trypanosomiasis Eradication Campaign aimed at eradicating tsetse flies; its chances of success are questionable, as is the wisdom of removing the threat of the disease from wide swathes of easily degraded rangeland (around one-third of Africa is denied to cattle ranchers), where it is effectively helping conserve the environment until workable sustainable ranching or agriculture can be developed.

In the Americas, from roughly the Rio Plata (Argentina) to Florida, a type of bed-bug transmits Chagas Disease (*Trypanosomiasis cruzii*), which causes serious debilitation and death. The spread is aided by the movements of organisms like house sparrows and by poor-quality buildings.

SCHISTOSOMIASIS

Although restricted to tropical regions shistosomiasis (bilharzia) has become more common in recent decades, probably because large irrigation schemes and reservoirs have favoured transmission. The use of more chemical fertilizers may also have reduced the acidity of some water bodies, favouring the breeding of the snails that harbour the flukes (fillaria) that cause the disease. There are three types of schistosomiasis, which debilitate and kill huge numbers of people in Africa, Asia, south east Asia and South America. Climatic warming may lead to more snail breeding simply because of higher temperatures, and wetter conditions also favour transmission.

RIFT VALLEY FEVER

This is a mosquito-borne viral infection, a cattle disease presently restricted to eastern Kenya and southern Somalia, which has jumped the species barrier to infect humans. First noted in 1931, it struck hundreds of humans, with a high mortality rate, in 1998 while there was flooding associated with an El Niño event.

HAEMORRHAGIC FEVER VIRUSES

In addition to dengue, already discussed, there are a number of viral infections that manifest as haemorrhagic fever. They include Hanta viruses, which are spread to humans through contact with the body waste of rodents (and possibly other animals like bats and fruit bats). In the early 1950s, over 2000 NATO troops in Korea were infected from field mice, which are also common in Japan, Russia and the Balkans. A less dangerous Hanta virus – Puumala virus – is spread in Europe by voles. Other parts of the world, including Brazil and North America, have Hanta viruses that are spread by rodents. They are likely to be affected by environmental changes such as adverse weather.

Also of cause for concern are the Filo viruses. There is some doubt about their mode of transmission, but no debate on how deadly the infections can be. There were outbreaks of Marburg Filo virus in Germany in 1967, and in Reston (USA) in 1989; in both cases the source was traced to laboratory monkeys. An especially deadly Filo virus is the Ebola Zaire virus, which causes severe haemorrhagic fever. An outbreak in Zaire in 1995 killed 92 per cent of victims and is highly infectious. These diseases seem to appear when people intrude into ecosystems where, presumably, there is an animal host. Environmental shocks like heavy rains or floods may bring host animals and humans into contact and then there can be an epidemic – rather like the pneumonic form of bubonic plague. These diseases are especially worrying because they can spread through human-to-human transmission with great ease once an outbreak starts.

Other haemorrhagic fever viruses include Flavi virus, Arena virus, Bunya virus (which includes the Hanta virus) and Nipa virus. Some outbreaks have been correlated with ENSO events (although this is unproven), which might favour rodent breeding (Grove and Chappell, 2000: 99). A threat is also posed by Sabla virus, Hendra virus, Guanarito virus, Australian bat Lyssa virus (in Europe, the spread of these viruses, which are related to rabies, is causing some concern) and many others.

UV RADIATION-RELATED ILLNESS

There has been growing interest in ultra-violet (UV) radiation-related health problems, mainly because of stratospheric ozone destruction by ozone-scavenging pollutants like CFCs. UV levels might increase for other, natural, reasons. Elevated UV levels can injure eyesight, causing cataracts, and promote cancer, especially skin cancer. Also, through disruption of human DNA, there may be altered immunity to virus, fungal and other diseases. UV radiation might alter the mutation rate of disease organisms. It might also affect livestock and crops, causing food shortages.

HIV/AIDS

HIV/AIDS is a major threat globally. Although subject of much disagreement, its origin may have been ape or monkey populations in newly cleared areas of Africa, from which the virus 'jumped' to humans. Once established in humans, the infection has been spread mainly by sexual transmission or contaminated needles and blood transfusion. The social and economic impacts are marked, especially in sub-Saharan African. By 2010 AIDS will have killed about 65 million people, putting it in the same league as Black Death and malaria (*British Medical Journal*, 324: 207, 2002).

5.16 El Niño and La Niña

El Niño

During 'normal' (non-El Niño) conditions, trade winds blow westward across the Pacific and 'pile up' warm surface water (see Chapter 3). Consequently, sea surface temperature is about 8°C higher in the western Pacific than off the west coast of South America. Cold water rising off South America favours rich plankton growth and the rainfall in the western Pacific is higher than it is toward South America. When an El Niño occurs, the trade winds ease in the central and western Pacific, waters cease to pile up and the west–east warm–cold anomaly weakens. The upwelling of cold water off South America declines, usually starting around Christmas, wildlife suffers and rains increase in western Latin America. At the same time in the western Pacific there is likely to be drought in Indonesia and Australia. These changes in patterns of surface water temperature have 'knock-on' effects on global weather patterns for months afterwards, the general effect being to bring drier conditions in southeast Africa, the western Pacific and Australasia, and northern South America, and to reduce the Indian monsoons (see http://www.pmel.noaa.gov/toga-tao/el-nino-story.html, accessed 25 May 2000).

The 1991–94 El Niño event coincided with severe flooding in southern

UK, bushfires in Australia, drought in northeast Brazil and Indonesia, heavy rains in Japan, and below-average rainfalls and droughts in southern Africa. An El Niño in 1997–98 caused economic disaster for Peruvian fisheries and adverse weather in California, western USA, Mexico and other parts of Central America. It probably triggered droughts in Australia, mild winters in eastern USA, droughts in the American Midwest, and ice storms in eastern Canada and New England (USA). Damages estimated at US$8 billion were caused (worldwide) by the 1982–83 El Niño event.

It must be stressed that El Niño impacts are still not fully understood, so accurate and reliable predictions are not possible (in the late 1990s Costa Rica moved cattle, expecting to have El Niño-related rains and these failed; livestock losses were considerable). Knowledge about the phenomenon is potentially valuable to forecasters. In Queensland, Australia, where farmers have successfully used software to factor El Niño into their long-term modelling and management, around 70 to 80 per cent of these farmers' profits are earned in the wettest three years of each decade; so being able to predict drought better is valuable (Couper-Johnston, 2000: 286). There is a further need for caution when applying historical lessons because El Niño events may strike modern society very differently to the ways they did in the past (see Chapter 4); local famines are less common and risks are spread much further afield and may nowadays affect distant economies (Couper-Johnston, 2000: 201). There are some indications that ENSO events have been becoming more persistent and intense over the past 80 years or so (perhaps due to global warming).

America's National Oceanic and Atmospheric Administration (NOAA) maintains a network of monitoring buoys across the Pacific, which transmit data that allows forecasters to give some warning months or even a year ahead of an El Niño. The TOPEX/Poseidon satellite, launched in 1992 by France and the USA, also collects valuable data that can be used for monitoring El Niño and La Niña events. El Niño forecasting predicts the behaviour of typical events, but not all are 'typical'. Reliable forecasting could help countries decide what combinations of crops to grow for the coming wet or dry conditions, prompt them to gather food reserves, make preparations for flood mitigation and landslide avoidance, etc. Peru, Australia, Brazil, Ethiopia and India have been exploring such forecasting, and a number of non-tropical countries like Japan and the USA are also developing it.

La Niña

A La Niña event (sometimes called El Viejo) is characterized by unusually cool ocean temperatures in the Equatorial Pacific (the opposite of an El Niño oscillation). The impact of La Niña seems to include raising winter temperatures in southeast USA, and depressing those in the North American northwest; there may also be altered incident, and shifted tracking of hurricanes in the Atlantic.

Key reading

Cartwright, F.F. and Biddis, M. (2001) *Diseases and History: the influence of disease in studying the great events of history.* Sutton, London.

Couper-Johnston, R. (2000) *El Niño: the weather phenomenon that changed the world.* Hodder & Stoughton, London.

Officer, C. and Page, J. (1998) *Tales of the Earth: paroxysms and perturbations of the blue planet.* Oxford University Press, New York (NY).

Smith, K. (1992) *Environmental Hazards: assessing the risk and reducing disaster.* Routledge, London.

Steel, D. (1995) *Rogue Asteroids and Doomsday Comets.* Wiley, New York (NY).

|6|

Reacting to environmental challenges and opportunities

... unfamiliarity of modern society with major natural catastrophes has bred general disinterest and complacency with regard to the threat we face from nature ...

(McGuire, 1999: 45)

Chapter summary

This chapter explores how humans react to environmental changes, addressing perception, early warning and adjustment to various hazards. There is a clear need to improve the awareness of the public, governments and planners to environmental vulnerability issues.

People are often ignorant of environmental challenges and are reluctant to invest in improving their knowledge. Identifying threats and opportunities is often far from straightforward, and after that it is necessary to improve the awareness of scientists, administrators and the public enough to make things happen. Even when a threat or opportunity is known to exist there is no guarantee any attempt will be made to respond in a rational way.

6.1 Perception

Human response to environmental change is affected by its frequency, severity and extent; past experience is important, as are the goals, attitudes, social pressures (and many other characteristics) of the people involved. Some environmental changes are 'instantaneous' and obvious, others 'creep up' and may be quite insidious (e.g. droughts, soil erosion, gradual climate changes), so people's awareness varies. Fear and response to change are often unpredictable: a group may dread something that presents much less risk than something they virtually ignore, and the effort to educate people about risks may be a challenging task.

There are three types of response to natural challenges:

1. do nothing
2. anticipate the change and try to prevent it, or (if possible) adapt to it and exploit opportunities and benefits
3. try to mitigate the change and exploit opportunities.

Most people see the Earth as dependable and stable – 'solid ground' – in reality it is really quite prone to change.

When disasters or catastrophes occur they are often regarded as 'acts of God' – unforeseeable events beyond human control and not worth planning for. Disasters are frequently seen as occasional events rather than ongoing problems (Whittow, 1980; Comfort *et al.*, 1999), and opportunities as provenance, occasional 'good luck'. A given challenge can prompt very different responses – for example, a storm can be greeted as 'God's judgement sent to test people' (the response perhaps being prayer for forgiveness), or it may be regarded as the result of witchcraft (the response might be to persecute witches), or it is seen as a natural event (the response may be to find ways to avoid future problems or mitigate impacts).

Most pre-modern societies were aware of threats and opportunities; they developed livelihoods and disaster avoidance or mitigation strategies that worked, or they failed to survive. Concern for natural hazards withered in recent times, reappearing mainly after the mid-1970s (White, 1974). Modern development has seen the degeneration or loss of many traditional livelihood and survival strategies, and too often something less secure and often of poorer quality has replaced these. It should be possible to add to the security and wellbeing of people, but only in the last few decades has there been concern for long-term sustainable development, with security a goal rather than the quick maximization of profits. Serious pursuit of sustainable development will demand more anticipation and preparation for environmental and social change and natural disaster.

Mainly since the 1950s there have been environmentalists keen to alert, or even frighten, the public and decision-makers into treating environmental issues seriously. But these Cassandras have often just catalogued threats and underemphasized the opportunities that may come from environmental changes. Environmentalists have launched into a lot of advocacy, sometimes with messianic fervour; but only comparatively recently has there been more development of monitoring, impact and risk assessment, and environmental management seeking well-thought-out strategies and objective measures. Governance and law-making are still adapting to environmental management demands.

A crude catalogue of recent environmental concern might be:

- in the 1970s – population growth and nuclear warfare
- in the 1980s to the 1990s – global warming, ozone destruction and loss of biodiversity
- from the 1990s to the present – global warming and genetic manipulation.

Less attention has been directed at natural climate variation, earthquakes, tsunamis, droughts and soil degradation. It is a little small-minded to focus on such oversights, though, because there has been tremendous progress in the last few decades in establishing real concern for environment–human interactions.

For some years there has been considerable interest and investment in anticipating, and preventing or preparing for, the effects of anthropogenic climate change (global warming). Some of these efforts have helped refine knowledge and techniques that are useful in dealing with many potential natural changes (Bruce *et al.*, 1996; Rayner and Malone, 1998). Hopefully, the world has now progressed beyond simply establishing that there are environmental threats, to actually seeking to be informed by environmental experts, and being willing to heed advice and pay to follow it (Nisbet, 1991; McGuire, 1999; Kasperson and Kasperson, 2001). There are a number of steps to be followed if there is to be a reduction of vulnerability to environmental threats and better use of opportunities. These are:

- realization that threats and opportunities exist
- perception that action is needed (may be improved by education and media efforts)
- understanding of the *causes* and development of effective countermeasures or adaptations
- willingness to spend money to find solutions
- governance, laws and administrative skills to successfully pursue and monitor effectiveness and, if need be, enforce solutions.

Unfortunately, some of the recent writing on environmental hazards has pandered to popular interest in 'conspiracy theories' and 'plausible disasters', and has been over-keen to argue predominantly natural causation for virtually any historical event. The risk is that a glut of this could lead administrators and citizenry to reject sensible warnings as unreliable neo-environmental determinism. More balanced and objective coverage of nature's interaction with humans is needed (see Glade *et al.*, 2001).

Planners have attempted to standardize terms: a *hazard* may be defined as a potential threat; a *risk* is the probability a hazard will occur. *Vulnerability* (see Chapter 4) is degree of sensitivity to, in terms of effect of impact, not awareness of hazard. For example, crossing the Atlantic by ocean liner or rowing boat incurs the same hazard: drowning. However, the risk of that fate is much greater in a rowing boat. A point already made is that, nowadays, it is often difficult to distinguish between natural and human-caused hazards because there are few aspects of the environment that are not in some way affected by humans and so no longer 'natural'.

Between 1970 and 1980 natural disasters killed over 3 million people, and badly affected over 820 million around the world (Smith, 1992: 22). Consequently, the world's media shows considerable interest in disasters – they are newsworthy. Yet people are often slow to accept the reality of a threat to themselves and to take precautionary action. In poor countries this is largely because most people are just too busy with day-to-day, hand-to-mouth survival; why threats are often ignored by richer people is less easy to explain. Various studies have highlighted the complexity and often apparent illogicality of hazard and disaster perceptions: real threats may be given less attention than unlikely, minor or even imagined threats. The media often adopt a distorted focus on developed-country cases, so that the death of a few westerners is likely to get far more media attention than hundreds of deaths in a developing country. The media tends to report the unusual (even bizarre) and the sudden. Disasters that are slow and gradual in onset tend to be overlooked or given less attention. There may be some slight justification for people and media focusing on the novel because, where disasters like hurricanes, floods or earthquakes are quite frequent (i.e. known), weaknesses in buildings and other infrastructure have probably been exposed already and people have made some adaptions.

Some environmental hazards are potentially so catastrophic that, although very rare, they pose a serious ongoing 'background' threat to human wellbeing. To those used to planning only a few years ahead the temptation is to dismiss them. Large volcanic eruptions, mega-tsunamis or extraterrestrial body impacts fall into this category. They recur quite often on a geological timescale and would cause vast damage and loss of life, but because they have not materialized during written history or human memory there is little fear of them. Palaeoecology and folk myths are often the only warnings we will get, and people may hesitate to be taxed, accept restrictive building regulations or evacuate a settlement because some administrator has deemed such evidence to be reliable. Sometimes predictions are imprecise or wrong, and heeding them wastes resources, upsets voters, and spells embarrassment for administration and forecasters. Reasoned argument, public education and careful lobbying of administrators are as vital as research into threats and opportunities.

The last few hundred years have been a relatively quiescent phase on Earth (which cannot last), history and palaeoecology warn us to prepare for less gentle conditions. McGuire (1999) tried to raise public awareness by sketching likely 'mega-catastrophes' he felt had a very real risk of occurrence within a century or so, which are well enough researched for scenario predictions to be reasonably accurate, and would have immense impact – at least civilization-threatening. These were as follows:

- *Volcanic super-eruption*: large caldera explosions and huge ash, gas and aerosol emissions. These would wreck national infrastructure, ruin the global economy and threaten harvests worldwide through 'nuclear winter'-type effects (sun blocking). Likely threat sites listed by McGuire (1999: 84) include Yellowstone (Wyoming, USA), the Cascade range (Washington State, USA), the Vesuvius/Naples region (Italy), Indonesia, Iceland, North Island (New Zealand), New Britain (Papua New Guinea) and Long Valley (California, USA), plus in all probability many other unsuspected sites. The Yellowstone region saw huge explosive eruptions as recently as roughly 2 million years BP and 660,000 BP.
- *Mega-tsunami*: the causes of these have been discussed in Chapter 5. Such a disaster would cause huge levels of damage and death in areas struck by such waves; the results would probably be the decimation of the global economy, possibly destruction of key food-producing areas, and might lead to warfare (the same is true of many of the other threats listed here).
- *Very severe earthquake affecting a major world city (e.g. Tokyo, San Francisco, or New York)*: the impact on the global economy would be severe.
- *Extraterrestrial body impact*: even quite a small body close to a city would have effects like those just listed above. Multiple strikes are a possibility. Larger objects might end civilization and endanger the survival of higher life forms.
- *Catastrophic human or crop disease outbreak*: there would be serious social and economic impacts, plus labour and skills shortages, but it would not automatically destroy infrastructure, nor the accumulated knowledge stored in libraries, and probably would not ruin agricultural land.

To prepare for such hazards decision-makers need to take the following precautions.

Pre-impact

1. Undertake pre-disaster planning. Avoidance planning and contingency planning.

2. Make preparations.

Dispersal of adequate emergency stores, preferably at safe locations and in secure, 'hardened' facilities. This is especially important for food reserves and crop seeds. Install early warning and avoidance systems; where there is dependency on a single city or nation for goods or services efforts should be made to reduce it. Try to ensure that dangerous activities, like nuclear waste disposal installations and nuclear power or weapons facilities, are protected against earthquakes, blast, tsunamis, etc.

Post-impact

3. Make appropriate responses after a disaster. This is aided by pre-disaster planning (aid, maintain public order, manage disaster).

4. Undertake effective recovery and reconstruction.

Precautions 1 and 2 are often neglected. During the Cold War most developed countries had some civil defence preparations, which would have served for coping with many natural disasters. Today, these have often been severely scaled down. Precautions 3 and 4 depend a great deal on 1 and 2 if they are to be effective.

The question of vulnerability was discussed in Chapter 4; one aspect of this is the greater exposure of poor people to hazards. Marxists might argue that reduction of hazard impacts depends on redistribution of wealth as well as improved warning and disaster planning. That is not the only viewpoint: the literature can be divided into that expressing a 'structuralist' viewpoint (human actions increase risks of natural disaster) and an 'underdevelopment/dependency' viewpoint (vulnerability is increased by poverty, marginalization and dependency). The richer nations cannot ignore the exposure of poor people to natural threats; they would face eco-refugees, epidemic disease and possibly warfare if disaster struck developing areas. Poverty and dependency alleviation should be addressed as part of disaster planning.

6.2 Early warning

Prediction merely states in advance the probability of a hazard occurring, it does not say when it will occur, nor what might actually happen. Forecasting involves assessment, perhaps well before the predicted event, but more often as things unfold. While developments like satellite remote sensing, improved telecommunications and GIS have helped, there is still a limit to what can be done without hindsight experience. Some hazards happen so infrequently that there is no hindsight experience available, so forecasting has to be founded on informed speculation and modelling. There may also be insufficient data on some threats to allow accurate predictions or assessments because the problem has only recently been recognized or adequate funding has not been made available to collect sufficient baseline data.

For many hazards there can only be warnings of risk and an estimate of the likelihood of occurrence. For natural threats that follow a predictable pattern it may be possible to monitor for critical thresholds and warning signs, and if these manifest take appropriate advance action. It is also important to structure the evaluation of threats to try and reduce the chances that something is overlooked. Impact assessment seeks to do this, but is not a precise, nor a wholly reliable, art.

Risk assessment aims to achieve the reliable identification and, ideally, quantitative measurement of significance and likelihood. There are two main categories of risk: *involuntary* (because they are unknown and/or unavoidable) and *voluntary* (known and accepted). These are not clear-cut categories: risks may be unknown because too little effort has been made to investigate and make people aware; voluntary risks may be accepted because there is no practical alternative, e.g. in poor countries people accept dangerous water supplies because they can afford no alternative). An 'acceptable risk' may merely be the least unacceptable option.

People do not make rational decisions; in particular, where there is a chance of profit, they often ignore risks. In addition, religious beliefs, fatalism or simple inertia may affect judgements ('God will provide or protect'; 'no rush; we will wait until the volcano really starts to rumble').

People often fail to heed a warning: the Captain of the *Titanic* appears to have been clearly advised of the risk of icebergs in the Atlantic! There are questions of legal liability and credibility involved in the forecasting and warning process: a warning that proves inaccurate might cause panic and costs; a few faulty warnings cause people to ignore the next, perhaps accurate, forecast (and to ignore even unrelated forecasts). Evacuation may leave possessions unprotected, so people may try to avoid moving, especially where they do not trust authorities to prevent looting. People can panic and be injured, riot, steal and behave in ways that may render them even more vulnerable. Those issuing warnings need to know how people are likely to respond, and what options they have that could reduce inappropriate behaviour and improve survival chances.

Hazard mapping is a tool that can aid the authorities to prepare for a disaster and to gradually reduce vulnerability in the most cost-effective way. Where the greatest hazard is mapped, people can be offered more training, emergency equipment and priority for evacuation; it may also be decided to increase taxes and control development to remove vulnerable buildings and activities from the threatened locality. Unfortunately, hazard mapping is not effective for all threats, although it is valuable when dealing with avalanche, mass movements, floods and bushfires. One must allow for the fact that successive disasters may have different impacts because the first removes the most vulnerable people and 'trains' others, or there may be progressive debilitation.

Early warning may mean minutes or years ahead. Where hazard maps have been published, mock exercises have been held, or people have been educated by pamphlets or briefings they are more likely to respond in appropriate ways. If early warning is short notice, such preparations are crucial. During the Cold War the USA used cinema films and school training sessions to train people in how to react to a nuclear attack early warning; in the UK the 'four-minute warning' might have sounded, but hardly anyone knew what it meant or how to respond!

6.3 Adjustment to hazards

Faced with a hazard, administrators and individuals can adopt a range of options between the two following extremes.

- *Accept the risk and do nothing* (because it is impossible to adjust, or the hazard has not been perceived, or the hazard is realized but dismissed as unlikely).
- *Manage all risks.*

The following strategies are available.

- *Modify the loss burden* – by means of insurance or some other form of loss sharing. The hazard is not reduced, but the impact is spread widely.
- *Modify the hazard* – try to suppress the hazard.
- *Modify human vulnerability to the hazard* – get people to adjust (through education, forecasting and warnings, land-use zoning, evacuation preparations and so on).
- *Disaster relief* – short- and longer-term aid after a disaster (survival aid, rehabilitation and reconstruction).

Insurance is essentially the spreading of impacts amongst many policy holders so that each bears a manageable burden (the premium paid each year). The problem with environmental impacts is that these are often widespread so that a lot of claims are generated, making it difficult to spread the risk as everyone is claiming at once. Insurers in the UK recently faced this situation: policy holders saw their 2001 premiums rise because of widespread and repeated flood damage in 2000. Insurance companies may spread risks by offering comprehensive policies to many people, not just specific flood-risk policies to a few; or by imposing excess charges (a pre-agreed sum is withheld from any claim); or by charging people in localities at greater risk higher premiums (loading a policy); or by capping pay-outs; by seeking reinsurance (the spreading of an insurance burden from a region to wider populations, even the entire world); or by excluding pay-out for certain risks (typically nuclear war or computer software error).

Insurance can be a problem where a large population is threatened, making it difficult to spread the risk adequately. There have been a number of huge claims on insurance companies in the last decade or so; for example, one US hurricane (in 1989) cost over US$15 billion, and one moderate and relatively localized Californian earthquake resulted in more than US$12 billion in claims (Kunreuther and Roth, 1998). When insurance companies are able to spread the risk adequately and offer cover, householders are notoriously reluctant to spend even small amounts, even when they are aware they are at risk. Since 1982 the French government has surcharged all insurance premiums to pay for a national hazard education programme.

In many countries there has been a reduction in disaster preparedness compared with the 1950s and 1960s, when there were functioning civil defence programmes that included food and shelter provision, and evacuation and emergency management preparations. These preparations were mainly intended to offer support in the event of nuclear warfare, but would be useful in the event of some other major disaster. With the end of the Cold War, countries like the UK cut back on civil defence. Some countries, like Switzerland, still require householders to install and maintain emergency shelters with supplies of food and water. It is common for people to be reluctant to fund and maintain hazard reduction unless there has been a recent and visible disaster.

Nowadays a good deal of disaster preparedness is undertaken by international agencies and non-governmental bodies like the Red Cross or Red Crescent organizations, and by smaller charities. For some hazards, as they are better understood, it may be possible to 'retrofit' to reduce vulnerability. That is, buildings can be strengthened to reduce earthquake damage, residential areas at risk of bushfires can be protected by firebreaks, coastlines vulnerable to tsunamis might have breakwaters installed.

6.4 Relatively well-known threats

Earthquakes

People in areas prone to frequent earthquakes are usually aware of the threat, but this is no guarantee that they will take reasonable precautions or respond in the best manner if one occurs. There is far less awareness of the threat in 'safe' regions well away from known 'quake areas, yet the risk is significant. Better awareness of 'quake risk should be promoted in areas where there has been mining, extraction of water or dissolved minerals through deep boreholes, waste disposal by pumping it down deep boreholes, areas where there has been nuclear weapons testing, and areas near or downstream of large reservoirs (all activities that can trigger 'quakes).

In The People's Republic of China, 'quakes killed an estimated 27 million and injured 76 million between 1949 and 1976 (Officer and Page, 1998: 36). The Chinese people and authorities treat earthquake threat seriously and have a well-developed monitoring and 'prediction' system. For the last quarter-century or so Japan and the USA have also been establishing monitoring and warning systems. However, even in these alert countries earthquakes still strike unexpectedly: a 'quake in 1989 at Loma Prieta (CA) collapsed the elevated Oakland Section of the San Francisco Bay Bridge without warning being issued; many were killed and injured in a location where there are probably more seismic measuring instruments in place than anywhere else in the world and structures are expected to be earthquake-resistant.

Those studying earthquakes have developed measures to help locate sources, assign magnitude and compare events. The point above the actual origin of a 'quake is called the focus and the actual point of origin is known as the epicentre. Sometimes less damaging foreshocks precede (and after-shocks may follow) the main shock(s); occasionally there is no warning before one or more severe shocks. There are two widely used measures of earthquake 'strength': the Richter Scale and the Mercalli Scale. The first of these, devised in 1935 by Charles Richter, is a complex logarithmic scale showing vibrational energy (established by observing the displacement of a seismograph trace and relating it to the distance of that instrument from the epicentre). It is an open-ended log scale, so point 4 is ten times as strong as point 3, and so on (Richter 8.9 is 700,000 times greater than Richter 5.0) (Box 6.1). The second widely used measure is the Mercalli Scale (Box 6.1). Nowadays, a 'Modified Mercalli Scale' is widely used, which expresses the *intensity* of a 'quake. It is a measure of the impact on humans, buildings, etc., and so is more a qualitative scale than the Richter Scale, which some claim is more 'scientific' (Table 6.1).

It must be noted that 'quakes measured to have exactly the same energy can have very different impacts, depending on many natural and human factors. Rock layers may focus shock waves – sometimes magnifying them several times. Unconsolidated sediments near the ground surface may liquefy and cause buildings to sink or slide, while those on more solid ground are undamaged. The depth at which a shock originates is important: shallow 'quakes (less than 40 km depth) are more damaging than an equivalent strength at greater depth. The construction type and height of buildings, the duration of the 'quake, the time of day and the preparedness of the people all very much influence the impact. Where a 'quake generates horizontal movement, rather than vertical, the damage is likely to be very

---

Box 6.1 Some earthquake disasters

- Shensi (AD 1556 – People's Republic of China) – probably over 800,000 died, many because they lived in loess-soil caves, which collapsed and buried them.
- Agadir (AD 1960 – Morocco) and San Francisco (AD 1989 – USA) – roughly the same strength earthquakes, but the former caused 14,000 deaths and the latter only 64. Similar strength 'quakes can have very different effects due to variation in geology, building styles, preparedness, time of day, etc.
- Haicheng (AD 1975 – People's Republic of China) – the city was evacuated when people observed increased flows from springs, and animals behaving strangely; consequently, in spite of a Richter 7 'quake, there was hardly any loss of life. Unfortunately, such warning signs are not always so apparent or reliable.
- Mexico City (AD 1985 – Mexico) – damage was varied, some taller buildings survived, while nearby lower-rise structures fell (because their natural resonant frequency matched that of the 'quake). Buildings on solid rock tended to fare better than those on softer lake sediments, which focused the shock and often 'liquefied' so that buildings simply sank and toppled.

Table 6.1 Modified Mercalli Scale of earthquake intensity with equivalent Richter Scale of earthquake magnitude

Modified Mercalli Scale			Richter Scale
	Manifest as:	Manifest as:	
I	Logged only by instruments	Not felt by people	——
II	Feeble	Felt by some sensitive people	——
III	Slight	Hanging objects swing	<4.2
IV	Moderate	Doors, windows and loose objects rattle, people notice even if walking	——
V	Slightly strong	Objects shaken, people woken	<4.8
VI	Strong	Windows break, walking is difficult, masonry cracks	<5.4
VII	Very strong	Difficult to stand, walls damaged badly	<6.1
VIII	Destructive	Weaker buildings collapse	——
IX	Ruinous	Houses collapse, ground cracking	<6.9
X	Disastrous	Widespread destruction of buildings, landslides, soil liquefaction	<7.3
XI	Very disastrous	Most buildings, bridges and infrastructure destroyed	<8.1
XII	Catastrophic	Total destruction, serious ground movement	>8.1 Richter Scale, in theory, runs to 10

Source: Based on various sources.

much greater. A number of disastrous earthquakes since AD 1900 have exceeded Richter 8.0 (Modified Mercalli Scale 'Catastrophic'), including: Colombia and Ecuador, 1906 (8.6); Chile, 1922 (8.3); Assam (India), 1950 (8.6); Chile, 1960 (8.3); Alaska, 1964 (8.4).

It is possible to draw up a list of 'at-risk cities and regions': Seattle, Vancouver, San Francisco, Tokyo, etc. These lie along the margins of the Earth's more active tectonic plates, with about two-thirds lying in the so-called 'Ring of Fire' around the Pacific. However, insufficient stress has been put on the fact that areas well away from plate margins, and that have been quiescent for decades or even throughout recorded history, can sometimes suffer catastrophic earthquakes. Examples of 'quakes away from plate edges and known surface faultlines include the one that hit Lisbon in 1775, and those that struck New Madrid (Missouri, USA) in 1811, 1812 and 1895. The latter registered as much as 8.7 on the Richter Scale, amongst the strongest recorded. The likelihood is that there will be a recurrence, yet people in the region are largely oblivious to the threat (McGuire, 1999). In the nineteenth century the population of New Madrid was sparse and lived in buildings of light construction, so there was limited loss of life; modern conditions would probably result in many more deaths. There were severe 'quakes in New York and Charleston (South Carolina) in 1886, and there could be recurrences; yet today there is virtually no popular conception of threat in these US cities.

Direct earthquake damage is only one part of seismic disaster: with gas and water mains and electrical cables torn, and cooking stoves upended, fire often rages; in steeper areas landslides, avalanches and mudslides are set in motion; in coastal areas or near lakes tsunamis may be generated, and there is a risk that large dams will be breached, causing catastrophic flooding (many have warned that the Three Gorges Dam in China poses such a threat). When risks are appreciated measures can often be taken to reduce earthquake damage; for example, Tokyo has installed computerized valves to cut off domestic gas supplies when there is a 'quake. Mitigating earthquake damage through building regulations must address both the problem of collapsing buildings and 'secondary impacts'. There are two approaches to building that can reduce 'quake damage.

1. Build structures so light that, if possible, they shake with little damage and do minimal harm if they do fall (e.g plastic tent structures, wood-frame buildings with lightweight roofs); although the structures must resist post-'quake fire and also have to withstand storm damage.
2. Reinforce and 'tune' stronger structures to likely 'quake frequencies. In addition it may be possible to mount buildings on shock-absorbing buffers and to place weights on appropriate floors to counter sway.

Resistant construction may give a false sense of security; if poorly consolidated sediment is built upon and 'liquefies' during a 'quake, even the most 'earthquake-proof' buildings can sink and topple. People may be prepared to trade off 'quake risk against building at a site where there is a spectacular view, convenient access to the workplace or cheap land prices – the poor have little or no choice, and are likely to live where there is tsunami and landslide risk. If 'quake-proof building is encouraged by regulations, or insurance companies refuse cover or raise premiums for at-risk areas and buildings, those who are less affluent and slow to respond may be left in property that has fallen in value and cannot be insured (or for which they cannot afford the premiums).

There is a possibility that some earthquakes might in future be avoided by countermeasures. These are far from fully developed, however there are potentially useful strategies: if a shallow fault is known to pose a threat it may be possible to pump fluid down a borehole to trigger a controlled 'quake or series of 'quakes, or simply to lubricate it so it can slide more gently; alternatively, an explosive charge might trigger a controlled 'quake. As these actions could be done after evacuation and with people alerted, the risks of a 'controlled' shock would be preferable to a natural occurrence, provided it actually reduces the dangers of the latter.

Earthquake forecasters face a dilemma: with imperfect methods they must decide whether to ignore a potential threat, or wait for it to develop more clearly (which reduces warning time and runs the risk that it will manifest before people are alerted), or risk a premature or false warning (Box 6.2). With reasonably accurate warning, utilities would be more willing to take costly ameliorative measures, such as reducing water levels in reservoirs (so dams are under less stress when there is a 'quake), evacuating people and halting gas supplies. Public evacuation is costly, presents problems for policing to ensure there is no looting and, if undertaken with undue haste and panic (a risk unless there is very good 'quake preparedness education backed by regular drills), may result in deaths, injuries and stress. If there are no workable evacuation plans and inadequate refuges it is probably better for 'quake victims to stay indoors, until after a 'quake, where they will be at less risk from falling debris (especially glass from high-rise buildings).

There is a risk of overlooking the cultural impact of natural disasters like earthquakes. As mentioned in Chapter 3, the Lisbon earthquake and associated tsunamis (AD 1775) not only destroyed one of Europe's major cultural centres, they also upset the complacent view most administrators and academics had about nature (Fig. 6.1).

Box 6.2 Earthquake early warning signs (not in order of promise or reliability)

- Monitor wildlife, domestic pets, and livestock – for unusual behaviour which may indicate approaching 'quake.
- Monitor for increased radon gas emissions in wells.
- Monitor water levels in wells.
- Use strain gauges – to detect strain build-up along moving faults, which start to 'stick' and generate strain prior to sudden energy release (the 'quake).
- Use levelling devices – to check for ground deformation.
- Monitor seismic noise – there may be a pause in the usual signals caused by a gradually moving fault, followed by a 'quake (a promising approach).
- Monitor gravity measurements.
- Desk research – establish the seismic history of an area.
- Research geological and extraterrestrial causes (tidal, planetary alignments) – in the hope of understanding the causes better.
- Monitor for static electricity activity (caused by earth movements).

Figure 6.1 The Lisbon earthquake, 1775
On 1 November 1775, Lisbon was shaken by an earthquake believed to have been stronger than 8.5 on the Richter Scale (the epicentre was roughly 200 km offshore). Building collapse, fires and three large tsunamis (which also struck Spain and Morocco, and were apparent in England) destroyed the city and killed roughly half of its population.
Reproduced with permission from the National Information Service for Earthquake Engineering, University of California, Berkeley (CA).

Tsunamis

As discussed in Chapter 5, tsunamis can be divided into 'normal' and 'mega', the former are roughly of 30 metres' height above sea-level (and may run up on to land to 70 m or more), the latter may be very much larger. Tsunamis travel at speeds close to that of a modern passenger aircraft, and an area may be struck once or repeatedly by a series of waves, tending to be more devastating where the sea gently shallows and the coastline reflects and concentrates the waves. Because tsunamis have long wavelengths they do not run up to give a brief deluge, but 'keep on coming' and cause tremendous damage. If settlement is sited where evacuation is hindered (e.g. by landward lagoons or cliffs) the impacts are likely to be more severe.

'Normal' tsunamis are common in the Pacific. Close to Japan's Sanriku Coast (Honshu Coast) the population and authorities take the risk seriously and invest in early warning systems, tsunami protection walls, offshore breakwaters, public education programmes and civil defence preparations, they plant coastal forest belts and site parks near the shore, and try to relocate settlement to higher ground. There are also high levels of awareness and some preparation in Hawaii.

Public education is vital. Without it warnings may be ignored or people may get impatient and return too soon to at-risk areas; there have even been cases where people who are unaware of the precursor phenomena flock to the shore to watch the sea recede, ignorant that it is a clear indication of a coming tsunami, which could have given them 30 minutes or more to evacuate.

The Pacific has had a tsunami early warning system (the International Tsunami Warning System) since 1948. It is operated by the US National Oceanographic and Atmospheric Administration (NOAA). This was initially intended to benefit Hawaii but is now part of a Pacific-wide system that uses a network of automatic tidal monitoring stations to identify unusual activity and transmit a warning possibly some hours in advance. Japan also has regional networks that give up to 20 minutes' early warning. With satellite remote sensing and automatic instrumentation that can transmit data from distant monitoring units tsunami early warning is improving, but is little used outside the Pacific. It is time nations bordering the Atlantic and other oceans took similar precautions.

There is far less awareness of mega-tsunamis amongst the scientific community (because convincing research was done only quite recently), and even less on the part of administrators and the public of developed and developing countries. Yet the evidence (see Chapter 5) points to frighteningly large waves. For example: the Alika slide (Mona Loa, Hawaii, *c.* 100,000 BP) generated waves that carried heavy coral blocks up on to islands 100 kilometres away to more than 375 metres above normal sea-level; El Hierro (Canary Islands) has suffered at least three similar collapses between 130,000 and 90,000 years ago; and on Eleuthera (Bahamas) huge boulders are stranded at 20 metres above present sea-level, perhaps testimony to tsunamis generated by one of the El Hierro slides. Apart from living well inland, and ideally at as high an altitude as possible, there is not a lot that can be done to avoid mega-tsunamis. At best there may be a few hours' early warning, which, without adequate advance training and contingency planning, would probably lead to chaos as great numbers tried to reach highland areas. Important resources and critical infrastructure could be relocated, duplicated or perhaps hardened. Many of the world's nuclear power stations and nuclear waste repositories are quite close to sea level and are not designed to resist wave damage or even flooding.

McGuire (1999) attempted to provoke interest leading to better awareness and preparation by presenting a fictional case study of a future mega-tsunami, based on the likely outcome if a huge landslide were to break free of the Cumbre Vieja volcano (La Palma, Canary Islands). There is already a fragment of the Island (*c.* 20 by 2 kilometres in size) that shows faultlines and subsidence caused by eruptions and earthquakes in 1941 and

1971. McGuire drew upon wave tank modelling, which predicts that a collapse would cause huge tsunamis (possibly well over 100 metres on reaching the USA and Caribbean); parts of Europe and Africa would also be struck by smaller, but still very damaging, waves. If effective monitoring is established, alerts issued from the Canary Islands might give several hours' early warning to the New World and a couple of hours to Europe and less to parts of Africa.

Volcanic activity

People are likely to be aware of the threat present in regions with active and recently active volcanoes. However, awareness seldom means satisfactory contingency planning and preparation; during the 2001 eruptions of Etna many households and businesses in at-risk areas had little or no insurance cover. Also, there are places with no known volcanoes where eruptions are possible, and others where the volcanoes are assumed 'extinct' because there is no memory of trouble. People are generally unaware that large eruptions can have very far-reaching effects: tropospheric ash clouds, noxious gas emissions and acid deposition can cause damage hundreds of kilometres from an eruption. Mega-tsunamis can be caused when volcanic islands blow apart, and particles blasted into the stratosphere could have worldwide 'dust veil' impact. Most commercial aircraft radar sets have great difficulty seeing ash clouds, so there is a real risk that jet engines could be snuffed out in flight.

There are situations where people's perceptions have proved woefully wrong:

- where a 'dormant' volcano has suddenly, and explosively, erupted
- an eruption venting down one flank of a volcano may suddenly shift to affect a different side
- ash clouds can shift with the wind to suddenly affect different areas
- people try to shelter from ashfall rather than evacuate, and then the character of the eruption changes to larger-size debris, pyroclastic flows, or the activity generates tsunamis or lahars; those surviving the ash may be killed by the latter phenomena.
- There have been situations where ashfalls have been unexpectedly widespread and damaging to agricultural production, infrastructure and communications (Box 6.3).

As with earthquake preparedness, public education greatly helps avoidance fatalities. For example, when Mount Pinatubo (Philippines, 1991) erupted, well-informed villagers in the vicinity evacuated with little delay, unlike those in other similar situations. This is attributed to the televising well in advance of video simulations of what was likely to happen. In 2000 Mexico's National Centre for Prevention of Disasters (CENAPRED) issued

---

Box 6.3 Some volcanic disasters

- Santorini (Island of Thera, c. 150 km north of Crete – various dates have been suggested, one possibility is 1628 BC). A series of eruptions and ashfalls might have allowed the Minoans on Crete to evacuate before a huge caldera explosion and tsunamis swept the island and the eastern Mediterranean (Friedrich, 2000). The impact on the Aegean was considerable and the Bronze Age Minoan civilization may well have been destroyed. Possibly this gave rise to the Atlantis legends and perhaps links with the events described in the biblical story of Moses and the Israelites' flight from Egypt (Exodus, Chapter 13: Verses 20–21; Chapter 14: Verses 21–28).
- Vesuvius (Bay of Naples, Italy, AD 79) – the settlements of Pompeii and Herculaneum, together with other towns, villages and scattered households were first deluged with ash and lapilli, and then struck by pyroclastic flows. Many who failed to evacuate were buried.
- Mount Pelée (Martinique, West Indies, AD 1902) – a sudden pyroclastic flow wiped out 29,000 people and left just four alive (of whom two survived their injuries) in the port of Saint Pierre. If people had evacuated *further away* they would probably have survived (warnings led to rural people moving into Saint Pierre).

---

two days' warning of an eruption of Popocatepetl, the Mexican Army evacuated thousands of people and the serious eruption caused no injuries. For busy, congested areas that are at risk, like Naples (Italy), early-warning instrumentation and evacuation preparations are important, although roads can quickly block if there are ashfalls.

People perceive volcanic activity as a 'far off' event; they are not aware that their climate and food supply or world trade could be hit even by a distant eruption. McGuire (1999: 64) pointed out that Europe has at least 41 active volcanoes and North America 185 (many in the Cascade range). Some of these volcanoes are serious potential threats on a regional, or even global, scale; Vancouver, parts of California and southern Italy are particularly threatened.

In the very distant past there may have been periods of much more intensive volcanic activity than now. Major climate changes involving sea-level variations and either the accumulation of thick 'ice age' snow and ice, or its melting, might increase volcanic activity (by altering the pressures on the Earth's crust). If this is true there may be periods with more eruptions over a timescale of a few hundred thousand years. The occurrence of mega-volcanoes has been discussed in Chapters 3 and 5; none exist at present but they might recur.

Where lavas are free-flowing (basic or basaltic lavas) eruptions are less likely to give rise to serious explosions, heavy ashfalls or pyroclastic flows – all likely where lavas are richer in silicon dioxide (acidic), which makes them 'sticky', and where there is a lot of superheated steam being emitted. The effect of an eruption is often influenced by weather and local topographic features: rainstorms may convert ash to dangerous mudflows (lahars), and eruptions may melt ice caps, divert rivers or breach lakes.

Lahars (see Chapter 5) have killed thousands in Java and in northern Andean Latin America. One of these disasters took place in Colombia in 1985, when a snow-capped volcano over 5000 metres in altitude erupted. The Nevado del Ruiz volcano caused lahars that swept into the town of Armero, killing around 22,000 people. Clear warnings had been issued in good time but people largely ignored them. There have been a number of studies of the perception of volcanic risk, willingness of authorities to issue warnings and public reaction (Metzger *et al.*, 1999).

The International Association of Volcanology and Chemistry of the Earth's Interior (Rome) maintains a Catalogue of Active Volcanoes, listing at present roughly 600, and about 1500 more or less active (i.e. that have erupted during the last 10,000 years). Smith (1992: 135) argued that, at the very least, volcanoes that have been active during the last 35,000 years should be treated as a potential threat. The number of eruptions each year currently averages around 50, most of these are located in the following places.

- *Where tectonic plates are being subducted* (e.g. along the western coast of the Americas). The process of subduction results in one plate being pulled down; as it plunges beneath the more stable plate it melts, and less dense material and water is incorporated into the molten rock, and rises. The resulting lava is acidic and gas-rich, which leads to often quite explosive and relatively unpredictable eruptions, where caldera formation and ejection of ash are more common. These volcanoes lie roughly along the edge of the non-subducted plate.
- *Where there is rifting*. This is the movement of plates apart (e.g. along the mid-Atlantic Ridge). The lava is usually basic and free-flowing, lessening the chance of explosive eruptions.
- *Stretching from above a hotspot* (e.g. Hawaii) there is typically a chain of islands that range from the newest, which is volcanically active (over the hotspot), with progressively older and less active islands or seamounts further away. This pattern reflects the passage of a plate past a 'hotspot' where there is probably an upward convection cell causing crustal rocks to melt. These tend to erupt basaltic lavas, which are less prone to explosive eruptions.

People tend to accept that there is volcanic activity, but assume that it will not happen in their time or particularly affect them. This is probably a reasonable attitude where eruptions mainly lead to limited ashfall and non-explosive lava flows – under such circumstances people can expect to evacuate after an eruption has started. There can be few countermeasures against really large explosive eruptions or very large fissure outpourings, other than monitoring for early warning and preparing effective evacuation arrangements for moving people some distance. Really serious eruptions would probably necessitate long-distance, perhaps inter-continental, evacuation, as well as permanent resettlement and serious disruption of food production and trade on a global scale.

For smaller eruptions it may be possible to halt or divert lava flows that threaten settlements or infrastructure, using armoured-cab bulldozers, by setting off explosions or by pumping cooling water over them. Buildings can be designed to better withstand or shed ashfalls (this may only require adding relatively cheap and simple steeply pitched roofs). High-risk areas can be mapped, and then residential development or important public works can be discouraged and evacuation plans prepared. At the time of writing (2001) less than 20 of the world's volcanoes were well enough researched and instrumented to offer reasonable early warning. Mount St Helens (USA, 1980) was monitored and warnings were issued, but people were still caught off-guard by the rapid and far-reaching lahars and sudden sideways-directed pyroclastic flows. In the case of really big explosive eruptions, there is a need for national *and* global preparedness, especially adequate emergency food stocks and world economic systems that are resilient even in the face of a major disaster. Planners, administrators and the public resist taxation and paying higher food prices to support the stockpiling of adequate emergency food reserves. The Mount St Helens eruption ejected enough ash to disrupt communications 700 km away and noticeably cooled the Northern Hemisphere for some months after; this was not a particularly large eruption, a really big one would seriously disrupt climate, possibly the stratospheric ozone layer, and thereby world food production and trade.

One of the most recent eruptions to have serious global effects was Toba (Indonesia, *c.* 73,000 BP). Although unproven, it might have helped initiate the last glacial. Tambora (Java, 1812 and 1815) was not especially large compared with some previous eruptions, yet in 1815 it killed 12,000 directly, and probably chilled the world from 1815–17, triggering crop failure and diseases, which carried off another 80,000 or more people. Past fissure-type eruptions were probably accompanied by sufficient outgassing and ash ejection to alter global climate and possibly the atmospheric gas mix, leading to mass extinctions. Nothing like these huge fissure eruptions has taken place during human history, although the Lakagigar (or Laki) fissure eruption (near Vainajökul in Iceland, AD 1783) formed lava flows of over 560 km^2 (14 km^3 of lava) in less than five months. And the gas and aerosol emissions – rich in poisonous fluorine, acidic sulphur dioxide and sulphides – caused crop- and livestock-damaging acid deposition and climatic cooling.

Mass movements

Wherever slopes exceed the angle for which the rocks comprising them are stable, there is a risk that gravity will cause a mass movement: landslide, mudslide, avalanche or rockslide. Snow is unstable on slopes of more than 20 degrees (unless anchored by vegetation); the critical slope angle for other materials depends on the character of the material and whether it is water-logged, faulted or disturbed. Landslides, mudslides and rockslides often take settlements by surprise because they have not happened in living memory. They may be triggered by unusual conditions: very heavy rain, earthquake or volcanic activity, or removal of vegetation by fire or over-grazing. Where mass movements are a regular occurrence people have often adapted. But sometimes people ignore history and build in vulnerable places.

Geomorphologists can effectively map mass-movement hazard, showing where it is safe to settle. Unfortunately, there are situations where the risk subtly changes without residents or specialists realizing: overgrazing may make movement more likely, ski resort development, changes in agricultural and forestry practice, global warming-related weather changes and acid deposition damage to forests may alter avalanche and rockslide incidence patterns and frequency. In the 1970s avalanches killed roughly 600 people a year around the world (Smith, 1992: 160). Although mapping and monitoring have improved, the increased development in at-risk areas has probably increased the annual death toll. In addition to hazard mapping, it is possible to engineer structures to hold unconsolidated material on-slope, or to slow, weaken or divert mass movement. Building regulations can discourage settlement and infrastructure development where it is at risk, and land use can be modified to reduce the risk (e.g. tree planting and pasture management to anchor soil and snow).

Extreme weather events

CYCLONIC TROPICAL STORMS

Cyclonic tropical storms (Atlantic hurricanes and their Australasian and Pacific equivalent – typhoons) usually cease to grow once over land, and die down as they track inland. There seem to be well-defined limits to storms on Earth. Very occasionally a cyclonic storm will linger in a position where it dies slowly and may even grow a little; for example, Hurricane Mitch, which struck Honduras in the late 1980s, 'settled' between a mountain range and the sea. Storm damage is caused by high winds and intense rain or hail, storm surges (raised sea or river levels) and storm waves whipped up by the winds, flying debris and lightning strikes.

Dust storms on Earth are limited in intensity, have a relatively restricted extent and short life thanks to the oceans, which prevent them growing. Dust storms can still have serious regional impact: in the 1930s people in the US Midwest suffered during the Dust Bowl disaster, and eye damage and serious respiratory distress were common. Large areas of sub-Saharan

Africa, North Africa, the Middle East, Asia and Australasia are prone to severe dust storms.

It will surprise many that the number of deaths from cyclonic storms exceeds that of all other frequently recurring natural hazards: fatalities are caused by the storms themselves, and by disease and famine due to damage to crops, and disrupted water supplies and infrastructure. Tropical cyclonic storms generally strike in known regions, although there may be occasional deviations from the average pattern (Fig. 6.2).

In 1938 New England (USA) was hit by one that wandered north, killing 2000, and in October 1987 southern England was hit by a category 1 (on the Saffir-Simpson Scale) hurricane (see Table 6.2). In 1991 a cyclonic storm hit Bangladesh and the storm surge it generated is estimated to have directly killed at least 200,000, possibly 400,000, people by drowning; disease, disruption of livelihoods and famine subsequently increased the death toll far more. One of the worst recorded Atlantic hurricanes was Hurricane Gilbert (in 1988). Luckily its track missed especially vulnerable areas. Lesser storms striking poor and crowded areas can cause far more death and misery; for example, in Bangladesh in 1970 over 300,000 were killed by a single cyclonic storm. The city of Galveston (Texas, USA) was hit by a catastrophic storm surge in 1900, which killed more than 8000 people.

Most developed countries track cyclonic storms and issue advance warnings; developing countries may not enjoy quite such effective early warning. In low-lying coastal areas, where there is only lightly constructed housing and no storm shelters have been provided, evacuation is the only effective option. People often hesitate to evacuate, and in regions like the Florida Keys (USA) people would have to move to safer ground via congested and vulnerable coastal highways (such as US Highway 1).

The number of hurricanes tracking across the Atlantic to the Caribbean and the USA seems to be increasing, up roughly 40 per cent compared with 30 years ago (McGuire, 1999: 34). This may reflect global warming and be an ongoing trend: hurricanes form where the sea surface temperature exceeds 26°C.

Table 6.2 The Saffir-Simpson Hurricane Scale

Scale number	Wind speed (ms⁻¹)	Surge (m)	Damage
1	33–42	1.2–1.6	minimal
2	43–49	1.7–2.5	moderate
3	50–58	2.6–3.8	extensive
4	59–69	3.9–5.5	extreme
5	>69	>5.5	catastrophic

Note: 'Surge' is the level to which the water level is raised by the wind.

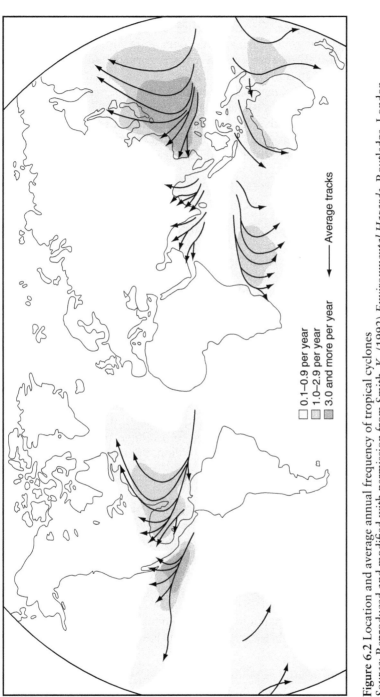

0.1–0.9 per year
1.0–2.9 per year
3.0 and more per year

—— Average tracks

Figure 6.2 Location and average annual frequency of tropical cyclones
Source: Reproduced and modified with permission from Smith, K. (1992) *Environmental Hazards.* Routledge, London.

TORNADOS

Tornados (see Chapter 5) often travel along relatively well-defined 'corridors', sometimes in swarms. Thus, damage is largely, but not wholly, limited to a relatively narrow swathe. A severe tornado will wreck virtually any building and can hurl heavy infrastructure some distance. In the USA, Texas, Oklahoma and Kansas are especially prone, but many other parts of the world can have the occasional tornado. Some US tornados have killed more than 700 people. Public fear of tornados is rather out of proportion to the actual risk, and in the areas of the USA most affected people often invest in storm shelters. Elsewhere, their appearance is a surprise and people are caught unawares.

HAIL STORMS, WIND STORMS AND ICE STORMS

Hail storms are feared by farmers in many regions. They strike suddenly and do great damage if crops are about to flower, or if grain or fruit are ripening. Attempts to predict hail storms and pre-empt damage by cloud-seeding over 'safe' areas has had limited success.

Wind storms can cause sudden damage to crops, forests, infrastructure and shipping.

Ice storms are a combination of weather conditions, which leads to severe icing. The cause is usually cold conditions, freezing mist or wet snow, and light wind. Power lines are brought down, trees and other vegetation are damaged by the sheer weight of ice, roads are made hazardous, ships may accumulate so much ice on their superstructure that they capsize, and bridges may be damaged. (Aircraft are rather less vulnerable in flight nowadays because they have anti-icing equipment.) Although often expensive, these events are not especially life-threatening; for example, in 1998 an ice storm in Canada and the USA cost over US$1 billion, but killed only 23 people.

TIDAL SURGES

Tidal surges can cause great loss of life, and damage to agriculture and infrastructure. At-risk areas are easy to map. However, the threat has largely been forgotten in southeastern UK and The Netherlands, even though in the late 1940s and early 1950s both were badly affected by a combination of high tides and winds, which backed up seawater, causing a number of deaths. London built the Thames Barrier to counter storm and tidal surges; however, many stretches of the UK's southern and eastern coastline have had too little spent on repair and improvement of coastal defences. Sea-level rises associated with global warming could cause havoc.

SNOW STORMS

Snow storms can cause economic chaos and deaths. The impact very much equates to preparedness. In Scandinavia, Russia, the European Alps,

Canada and continental USA, they happen often enough that people are experienced and authorities make preparations. Sometimes the storms are especially severe or strike normally unaffected areas, and then people and governments are caught unprepared. Forecasting can help, but may be ignored, and cases of exposure, heart attack and motor accidents strike the unaccustomed. These fatalities might be reduced by public education, and building regulations or government subsidies can help reduce structural damage and the effects of cold.

ABNORMAL FROST

Abnormal frost may take a number of forms. Severe frost might catch people reasonably used to cold conditions off-guard, causing deaths by exposure, road accidents, communications chaos and building damage. However, more impacts are likely to be felt when frosts occur in parts of the world that are seldom affected by them (e.g. Florida, the Papua New Guinea Highlands, southern Brazil). Unseasonable frosts can also catch people unawares even where winter chills are common. Where rogue frosts are a regular threat, people usually grow crops of a type less likely to suffer, in places less affected or in ways to lessen impact. In Florida, the response of growers to severe frost damage in the 1960s was to rethink what they planted, favouring more robust or earlier-maturing but less profitable varieties, and to shift planting where possible to be close to water or further south.

Temperate USA and Europe have records of 'freak' frosts, even in lowland areas, at midsummer (on average once every few hundred years). Frosts may cause damage by striking just at a crucial time: when fruit trees are in flower; and the night before the ill-fated Spaceshuttle *Challenger* launch (chilled seals leaked hot gases and helped cause the explosion).

Floods

Some floods are generated by sudden intense rainfall or rapid snowmelt, others result from prolonged gentle precipitation, tsunamis, tidal surges, obstruction of a channel by debris or ice, and overgrazing or bushfire, which removes vegetation from a watershed promoting flash-flooding. Along the Mississippi and Missouri (USA) in spring 1993 there were severe floods, and in the UK in 2000/01. Floods along the Rhine in 1994 necessitated the temporary evacuation of 92,000 people, and further floods in 1995 caused Dutch authorities to temporarily evacuate about 85,000. In densely settled and low-lying countries like Bangladesh storm surges and river floods can easily affect millions. It is suspected that El Niño events may help trigger flooding in a number of countries in the Americas, Africa and Australasia (for example, the flooding in Mozambique in 2000). Indigenous people may well have had strategies to cope with such events, but these have increasingly been abandoned, and there is a need to find alternatives that work under modern social and economic conditions.

There is insufficient awareness of the importance of flooding: a large proportion of the annual victims of natural disasters suffer through floods. For example, in 1991 and 1998, Chinese flood victims accounted for more than half of the world's natural disaster total (International Federation of Red Cross and Red Crescent Societies, 2001). In developed countries the favoured adaption nowadays is insurance; however, many vulnerable people fail to take out policies, either because they do not perceive the risk, feel they cannot afford it or simply cannot be bothered. In the UK since 2001, government and insurers have provided adverts, free telephone helplines and websites to advise people whether they are at risk from flood. There has also been an improvement of early warning in the form of broadcast flood alerts. Flash floods strike vulnerable areas suddenly, possibly at night when people are asleep. The cause can be rainfall in uplands far away from the lowland areas struck by flooding, the likelihood is that it is not even raining in vulnerable areas, and when this is the case telephoned or radio warnings may be able to give considerable early warning.

Drought

Drought and desertification have been vaguely defined and this has resulted in often inaccurate and unreliable statistics. There has been debate over what is responsible for drought and the associated ills of famine and dryland degradation (desertification). Some drought may be due to natural causes, but there are situations where human mismanagement is to blame and some where it is a combination of both. Where natural changes are involved, the expenditure of funds and attempts to control land use may be largely a waste, and the area might be better abandoned. Sometimes drought threat may be reduced by careful land management. People's traditional adaptations to the threat of drought (see Box 7.2) have often been degraded by modern development pressures.

Davis (2001) recently provoked debate on droughts after publishing his 'political ecology of famine' views. He argued that 'climatic accidents' – notably the 1875/76 great drought that hit northern China, North Africa and other countries depending on the tropical monsoons, and also similar droughts in 1896–1901, and in 1877/78 in Morocco, Brazil and India – were better withstood before colonial intervention. Davis held that economic changes and loss of sovereignty had much to do with these 'late Victorian holocausts', and warned that present-day globalization and other trends may again be making the world as vulnerable as some colonies were in the 1870s to early 1900s.

The Sahel Zone of sub-Saharan Africa receives its precipitation as a consequence of seasonal movements of the Intertropical Convergence Zone (ITCZ): northeast trade winds carry dry air across the Sahara; to the south are the southeast trades carrying moist air from Central Africa (for one to four months each year the ITCZ loops northward bringing rain to the Sahel). This 'monsoon' is variable in length, intensity of precipitation and the areas it affects, so droughts are a threat. The movement of the ITCZ seems to be influenced by ENSO events (see Chapter 5), so there is considerable hope of better warnings of drought because the initial development of an El Niño precedes the start of drought over Africa by several months. Unfortunately, the reliability of such forecasts is limited.

Even if forecasting remains imprecise, areas badly degraded by drought and desertification might be rehabilitated during wet El Niño phases. If these can be identified reliably in advance, and the way they develop can better be understood, it may be possible to remove livestock, and the above average precipitation would then allow otherwise unlikely vegetation and soil recovery.

Key reading

Kasperson, J.X. and Kasperson, R.E. (eds) (2001) *Global Environmental Risk*. Earthscan, London.

McGuire, B. (1999) *Apocalypse: a natural history of global disasters*. Cassell & Co, London.

Officer, C. and Page, J. (1998) *Tales of the Earth: paroxysms and perturbations of the blue planet*. Oxford University Press, New York (NY).

Smith, K. (1992) *Environmental Hazards: assessing the risk and reducing disaster*. Routledge, London.

|7|

Coping with nature

We are the offspring of history, and must establish our own paths in this most diverse and interesting of conceivable universes – one indifferent to our suffering, and therefore offering us maximal freedom to thrive, or to fail, in our own chosen way.

(Gould, 1989: 323)

Chapter summary

This penultimate chapter explores responses to environmental changes that seek to reduce unwanted impacts and to exploit opportunities. In the past people developed coping strategies; modern humans need to expand theirs. Failure to respond in the right way will result in problems like the dislocation of large numbers of eco-refugees.

Burton *et al.* (1978) have explored how individuals and social groups respond to extreme events in nature. They stressed that there are often benefits associated with change, even if risks may increase, and that the negative aspects are usually recognized and the positive overlooked. Accurate estimates of world hazards are hard to come by, and those of Burton *et al.* (1978: 2) are both approximate and now rather dated. They estimated that in the decades prior to 1978, 90 per cent of natural disasters were due to four types of hazard – in rank order: floods (most frequent and most damaging to property); tropical cyclonic storms (which cause most fatalities); earthquakes; drought. Also, they suggested that about two-thirds of natural disaster-caused deaths were in developing countries.

Human activities are altering natural processes, often in subtle ways that alter the scale and speed of impacts. Effective environmental management demands better understanding of environment–human relationships (McMichael, 2001).

7.1 Responding to environmental changes

Adaptation and adaptability are the processes whereby beneficial relationships between humans and their environment are established and maintained, and 'allow human populations to change in response to changing or changed environments' (Ulijaszek and Huss-Ashmore, 1997: 7). Adaption is a process of change; an adaptation is a state. In the process of adaption, humans may alter the environment, and vice-versa; it is a two-way process (Dubos, 1965). Adaption is less easy when there is sudden and rapid change. A wide variety of academic disciplines have focused on adaption, including: genetics (via selection); physiology (looking at short-term, individual responses to factors like cold, altitude, changing food supply and so on); behavioural studies (examining behaviour changes that give an advantage); and cultural studies (focusing on transmission of ideas, knowledge and technology) (Cullins and Weiner, 1977).

Adaptive processes do not operate in isolation, there are feedbacks and linkages between different things; some adaptive processes are rapid, others slow and patchy (e.g. the shift from nomadic hunter-gatherer to modern human society is still going on). Not all adaptions are useful (at least at present), many westerners are becoming obese in response to an inactive lifestyle; in the past, in times of food shortage, this may have been beneficial, now it is life-threatening. Research on human adaptation was expanded between the 1950s and 1980s by the International Biological Programme, the Human Adaptability Programme and the UNESCO/UNEP Man and Biosphere Programme.

Response to natural change is affected by people's perception of what is happening and their awareness of opportunities to adjust. Responses can be short term (e.g. rapid cultural changes), long term or very long term. The latter may arise through mutation so that people can function better: e.g. the sickle-cell trait, developed by mutation, gives better resistance to malaria and gradually spreads through an at-risk population.

Responses to threats can be difficult to predict. Adaption may not always be satisfactory – in northeast Brazil the peasantry may survive drought but through a life of poverty – there can be situations where adaption has 'seized up' and people somehow fail to progress to a new state of reasonable well-being. This may simply be that the struggle to adapt has used up all the available resources or human willpower. Environmental knowledge can mean different things to various social groups (Eden, 1998). Some may find the opening quotation at the start of this chapter offensive: Gould clearly charges humankind with control of their destiny, with no supernatural intervention envisaged.

Nowadays, the 'visibility' of a problem is in large part determined by media coverage, and the more visible it makes a threat or opportunity, the more likely it is that people will demand action or respond themselves. Burton *et al.* (1978: 137) noted that the public tend to have a limited attention span – news soon becomes 'stale' and something new is presented; media developments in the West since the late 1970s may have caused these changes. It is also important to note that special interest groups and elites have considerable influence over what adaptions are made.

Recognizing and warning of a threat is no guarantee of a satisfactory response. For example, the 1970 Bangladesh cyclone disaster (which killed roughly 225,000 in its first 12 hours) was tracked by meteorologists, giving ample time for radio broadcast warnings. However, little was done because the cyclone was not considered especially severe. Unfortunately, it coincided with high tides. The Mozambique flooding in 2000 was predicted using knowledge about El Niño events and warnings were issued to the press and various agencies as much as six months ahead (Pearce, 2000b), unfortunately relatively little action was taken to mitigate the impacts.

People *may* cope better if they are used to a hazard compared with being suddenly exposed to something unfamiliar. There are some situations where people are quite familiar with a threat, yet fail to act wisely. The impact of a hazard may be a function of a number of factors, not just severity, and of how it coincides with other events, its timing, duration, etc.

Western society often believes it is in control of nature; the reality is more a sort of 'ecological hubris' – overconfidence and failure to think (Sale, 1985: 15). If this attitude is not corrected it can lead to catastrophe. Even in the last 150 years there have been several disasters that could have been avoided; one of these was the Dust Bowl in the US Midwest in the 1930s, which caused soil degradation, poverty, illness and the mass exodus of 'eco-refugees', most headed for California. An earlier failure to avoid disaster was the 1845–47 Irish Potato Famine. The human population had climbed from less than 2 million to over 8 million, and became too dependent on the potato crop, which had little resilience when disease struck. The numerous vulnerable tenant small-farmer families were quickly destitute and hungry. In 1845 as many as 4 million died or were forced to emigrate, and many survivors still suffered poverty a century later. Land reform and crop diversification could have helped avert famine. There are other examples of human failure to avoid disaster – Easter Island is often singled out as a case study (Hamaker, 1982).

Humans today must worry about both natural threats *and* anthropogenic threats. There is a lot to worry about and it might be easier if done as much as possible at a manageable local or regional scale (with global-scale coordination, aid in time of disaster, monitoring and research). Sale (1985) argued for the bioregion as an appropriate unit; similar suggestions were made by Berg and Dasmann in the 1970s for coping with problems and for seeking secure, sustainable development. Some oppose such proposals, fearing 'balkanization' (the assumption that it will prompt separatist trends). As stressed in Chapter 4, people today are probably far less adaptable than at any point in the past; breakdown is a distinct possibility if a disaster disrupts food production, water supplies or governance. *Wherever possible, measures should be taken to increase human adaptability and resilience.* Unfortunately, some aspects of globalization seem to be prompting even more uniformity of consumption, production and dependency on worldwide and vulnerable economic systems and communications. Need this be the case, or can globalization be 'tuned'?

Apparently entrenched adaption to risks can rapidly change – for example, when people see an advantage in an innovation. There is also considerable variation in the amount of stress people can take before changing their ways or giving up. There have been suggestions that urban people are increasingly less aware of nature, which given their 'artificial' environment may often be the case. Over half of the world population are now 'urbanites' and their numbers are growing; they are less likely to make adaptions and may actively oppose actions and expenditure that favour non-urban areas and peoples, yet they depend on these for their food and water (and often energy).

Coping strategies

These have been defined as short-term or temporary responses to external shocks and stresses. Depending on the severity and length of the shock, coping strategies fade away and normal livelihood strategies return (de Haan, 2000: 348). Sometimes coping strategies become the norm and can be said to have become adaptive strategies. There can be individual responses and collective responses to environmental change and hazards (see Boxes 7.1 and 7.2). In Alpine Europe householders tend in the main to

---

Box 7.1 Options that may be available as a way to cope with environmental change or hazards

- *Innovation* – find alternative ways of making a living.
- *Cooperation* – grouping together with others for mutual support. The establishment of cooperatives can be a powerful way of providing insurance, and supporting research and innovation.
- *Diversification* – adopt more than one livelihood element, ideally with different vulnerability; for a farmer this means a greater variety of crops or adopting non-agricultural activities, e.g. on-farm tourism, growing narcotics on common land, etc.
- *Insurance* – at present mainly restricted to developed countries (spreads the burden of coping among a wide group of the population); schemes often need to be supported by some form of re-insurance.
- *Intensification* – produce more from available land if other is unavailable.
- *Empowerment* – in recent years this has increasingly been advocated, the idea being that people have more autonomy and control over their fate and so are less vulnerable.

Box 7.2 Some adaptations a peasant farmer may select when faced with drought

- Offer the farmer's (and perhaps his family's) service as waged labour.
- Make things for sale – wood carvings, mat weaving, etc.
- Seek alternative foods/forage – wild food sources that are largely ignored in better times because they are inferior to crops in some way.
- Turn to crime – banditry, smuggling, poaching or, if there is land and enough moisture, grow narcotics.
- Mobility – migrate to find off-farm employment; move to friends or relatives in a different area. This may involve costs and be hindered by the need for travel and employment documents, border and other controls.
- Cut firewood or make charcoal for sale.
- Sell some livestock, family possessions, land or timber – although at a time of stress the market may be glutted and purchasers few and far between.
- Innovate – unlikely in time of stress, but not impossible.
- Seek credit – few are willing to lend to high-risk candidates at a time of stress and uncertainty.

Note: Modern pressures have often disrupted the above adaptations.

---

leave avalanche adaptions to their community, but may make personal adaptions to other threats and opportunities. Some African countries have laws requiring farmers to plant a given area of 'famine foods' every year in case of drought or other cause of crop loss. If a government wishes to encourage individual householders to prepare for natural challenges it has a number of ploys it can use – legislation, the offer of grants, subsidies or tax allowances – or it can seek to educate them.

Coping with environmental challenges often seems to be based on decisions that do not really consider utility (Burton *et al.*, 1978: 85–91). Mortimore (1989: 3), exploring adaption to drought, observed that individuals, families or communities have to confront the situation in which they find themselves at a given time, with the resources they are aware of, under given constraints of land, labour, capital and mobility. The process of longer-term adaption (adaptation) is both biological and cultural (for a fuller discussion see Bennett, 1976b). Adjustment is the incidental and purposeful short-term reaction to a challenge (steps en-route to adaption). It is probable that much adaption is ad hoc and undertaken year on year. The rich should find it easier to adapt because they have better resources to experiment with, to support a change of livelihood or a relocation.

Burton *et al.* (1978: 29) recognized two categories of hazard: 'pervasive' and 'intensive'. The former are relatively slow to develop (which means there is more chance they will be identified and even avoided, or their impact reduced) and, because they are often perceived to be longer-lived threats, are more likely to prompt adaption and spending on prevention or mitigation (e.g drought). The latter are sudden, and because people are often taken unawares, less likely to prompt preventative or mitigation measures (e.g. tornadoes). In the USA droughts kill few people nowadays and tornadoes typically kill more than a hundred people a year, yet the latter attracts less spending.

Box 7.2 lists some adaptions a farmer or herder might make when faced by drought. For those who are part of the cash economy and with enough surplus income, if carefully planned insurance can be a way of helping spread the impact of natural impacts and of encouraging changes in behaviour to reduce the risks (i.e. avoidance as well as relief). It is crucial that insurance is adequately underwritten so that widespread claims do not cause the system to collapse. Those taking insurance can be 'encouraged' by premium incentives and conditions on how payment is made to behave in ways that reduce vulnerability; for example, relocation to safer sites can be encouraged by demanding high insurance charges in risky areas, and reductions can reward those who own more sturdy buildings. The very poor and those in subsistence economies find it difficult to purchase insurance; schemes based on sale of surplus livestock, or crops in times of plenty and emergency food-for-work schemes have been tried in a number of countries.

Avoidance strategies

As technology develops the chance of avoiding or mitigating natural hazards will hopefully improve. Traditionally, societies often developed hazard avoidance or mitigation strategies (an example of this – 'potlatch' – was mentioned in Chapter 3). Many countries seek to reduce the chances of bushfires by pre-emptive burning, creation of firebreaks (belts of land cleared of flammable vegetation), firewatch and early warning so that outbreaks can be tackled before they get unmanageable. There are river basin authorities that seek to coordinate and manage streamflow (through dams, barrages, flood relief channels and diversion of water on to flood-control areas) and run-off from the land in order to reduce flood damage. Improved communications aid the transmission of warnings, and the movement of experts and aid.

Avoidance may take a lot of research and a lot of money and, in most cases, demands good preparation and coordination. There have been hopes that severe storms might be steered away from populous areas or weakened; since 1947 cloud-seeding with silver iodide has been tried (the seeding should provide freezing nuclei, prompting the supercooled moisture in clouds to freeze; the release of latent heat involved will then hopefully reduce the wind speed, and the risk of intense rainfall over a vulnerable area is reduced). However, it has so far given inconclusive and even worrying results. Experiments with cloud-seeding to weaken hurricanes have been undertaken by the USA (in 1961, 1963, 1969 and 1971) and other countries, but the real value of cloud-seeding is still unclear. There is some military potential in being able to steer and vary the intensity of storms, so some research may be unavailable for civil use.

Until recently most efforts at avoiding natural hazards were based on technological measures, now there is growing interest in social adjustments and in ensuring that technology is appropriate and less likely to fail during disasters. An example has been the promotion and spread of stonelines in West Africa; these are structures formed of locally available rocks laid along contours to help soil and moisture conservation, and so reduce the risk of crop loss when rainfall is scarce or torrential. Improved yields help ensure that people are healthier and have more resources to withstand difficulties. The lines can actually collect debris and so make sustainable production possible (Atampugre, 1993). A point made by Burton *et al.*, (1978: 172) is that it is probably easier and more attractive for countries to collaborate (especially when poorer countries are involved) on avoiding or mitigating natural disasters than on vague environmental protection and conservation, because it is seen to be 'relevant' and beneficial to humans.

7.2 When response to environmental change is inadequate

The challenge is for people to make sensible development decisions in an uncertain world. Governments vary a lot in the support they offer people to help them cope with natural problems. There may be considerable inequity; recently in the UK there has been some compensation for livestock lost to foot and mouth disease, yet farmers losing crops through insect pests, frosts or flood get little help, and other rural livelihoods like the tourist industry have had limited aid. By and large it is larger disasters that get attention and attract compensation. Self-help, like the stonelines just mentioned, can ensure that there is minimal dependency – making local people responsible for their own security – and often allows a rapid response.

Eco-refugees

Use of the expression 'eco-refugee' (or environmental refugee) can be traced back to El-Hinnawi (1985), who applied it to those forced to flee from environmental problems (natural changes, environmental disasters *and* anthropogenic disruption of nature). There have been other attempts to define the term; for example, 'people forced to leave their traditional habitat temporarily or permanently because of environmental disruption (natural or human caused)'. People may fail to withstand a natural challenge without moving; eco-refugees may make their move gradually, or they may suddenly be forced to relocate en-masse. Some may be temporary refugees, others are more permanently excluded from their homes.

Eco-refugee is not a precise, nor a wholly satisfactory term; it has legal and conceptual weaknesses (for an overview of refugee issues see Harding, 2000). It is sometimes difficult to recognize true causes for the movement of people – there may be a mix of social, economic and environmental – indeed, there is often no simple and direct cause-and-effect link between environmental change and relocation. Sometimes environmental pressures help trigger a political or social conflict, which then generates refugees: Ramlogam (1996) argued that Rwanda's genocide in the 1990s, the Vietnamese 'Boat People' and those fleeing Haiti were at least part-eco-refugees, encouraged to move by land degradation or bad weather.

A given environmental change might bring into play effective adaptions (often based on use of social capital) in one community, which fail to appear in another, so that the impacts are very different. A sudden natural disaster may cause many or a few to flee immediately, or relocation could be a much slower (even almost insidious) process in response to gradual deterioration. People may change their attitude and be more or less pre-pared to endure hardship (even if conditions have not deteriorated much), or a situation may have slowly degenerated until a 'threshold' is reached beyond which relocation takes off rapidly. Eco-refugees are generated when people elect or are forced to move, but this is just one of many coping strategies. Causes are seldom simple and clear-cut: drought may or may not cause famine and when it does there are often social and economic factors involved.

Some eco-refugees move within their country, others cross borders. This is nothing new, humans have always used movement as one of their main adaptive measures; what has changed is the possible scale of the problem with modern cities often housing several million, and crowded lowlands and coastlands perhaps tens of millions. In the 1840s, those people who could afford it fled the Irish Potato Famine, joining a flood of other European migrants who had been settling the New World, Australia, South Africa and the Pacific for 200 years or more. An unknown quantity is to what degree European emigration after 1650 (over 50 million between 1820 and 1930) was due to the Little Ice Age (it was probably at least in part a flow of eco-migrants). European migration was certainly triggered by various forms of persecution, opportunism, population growth and improved transporta-tion, but the economic hardships were probably partly attributable to environmental factors (Crosby, 1986: 5).

By the 1940s most frontiers had 'closed', i.e. there was little unoccupied and usable land still accessible and attractive to settlers. Eco-refugees since the 1960s have mainly been from Africa and Asia to Europe and North America. Establishing to what extent environmental factors have played a part in these movements of people is difficult, many migrants have come from urban areas, but some can trace a path from rural farming roots that they have torn up because a combination of developmental and environmental factors made their agriculture inadequate or unviable.

Eco-refugees can be divided into:

- those shifting because of population pressure and marginalization often associated with land degradation
- those disrupted by global change, likely to be linked mainly to greenhouse effect warming
- victims of natural disasters.

In 1971 the FAO established the Global Information and Early Warning System (GIEWS); its principal objective is to monitor and warn of food supply problems. In the three decades that followed there seems to have been limited progress in terms of warning of situations that generate food-supply problems and refugees. Some developed countries are becoming concerned at the possibility of large numbers of eco-refugees who may seek to cross their borders, and many developing countries are keen to reduce rural–urban migration, so there is growing interest in the factors that trigger the movements for example, in the Maghreb (Africa, north of the Sahara), where land degradation could drive large numbers to migrate to urban areas that can offer little employment. Developing countries are also becoming aware of the costs and problems they may face if large numbers of eco-refugees arrive from neighbouring countries or marginal areas; land around resettlement camps may get deforested, soil and water supplies may be over-stretched to the point of breakdown, and there may be socio-economic impacts on the host population and conflict with them (Sudan had such problems after famines in Ethiopia in the early 1980s caused refugees). Döös (1997: 49), seeking to predict likely scenarios for 1990–2030, warned that past flows of eco-refugees are of little or no help in forecasting future movements.

Refugees – eco-refugees included – usually have little capacity or motivation to maintain their new environment in good condition; they are likely to be traumatized, have often lost their money, tools and livestock, may be suffering malnutrition, and remain unsettled and insecure. Host populations may resent the influx, even when, like the eco-refugees fleeing the US Dust Bowl of the 1930s to resettle in California, they are of the same race and nationality. Having probably found that their normal coping responses are useless in the new area, refugees are unlikely to be familiar with the dangers and opportunities of their new surroundings. Unless carefully managed, assistance may engender jealousy from the host population, who feel that strangers are getting aid they have never enjoyed.

Debilitated people moving to new environments can introduce diseases that may encounter little resistance and cause serious epidemics among the host population as well as refugees.

It can be difficult to be really sure what 'pushes' and what 'pulls' refugees into moving; the spread of TV, better road and air communications, and the stories told by past labour migrants may play as much part as deteriorating environmental conditions. As discussed earlier, land degradation and population increase might make people more vulnerable to natural environmental changes, but may also prompt intensification and a move toward sustainable livelihoods.

Myers and Kent (1995) estimated that, worldwide, there were at least 25 million eco-refugees, compared with about 23 million fleeing economic and social causes. The total might double by 2010, through ongoing land degradation, even if there is no significant environmental change. However, if sea-levels rise a metre or so, there could be 200 million or more. Of the estimated 25 million eco-refugees in 1995, roughly 5 million were in the African Sahel and 4 million in the Horn of Africa. Haiti may have lost one in five of its population through environmental problems and political unrest, the main destination being Florida (USA), where the consequent burden has been marked (Myers, in Munn 2002: 214–18). Munn (2002: 414–18) and Westing (1992) also predict a marked increase in eco-refugees, especially in Africa, and call for aid to be targeted at the root causes.

In recent years there has been growing awareness that future global warming may generate eco-refugees in large numbers. Some of the interest is possibly prompted by fears that the international community will have to assist (Myers, 1993; Black, 1994; McGregor, 1994). The focus has been mainly on populous low-lying countries and areas – Bangladesh, Egypt, the Maldives, Kiribati and eastern USA – which would be seriously threatened by rising sea-levels. Possible sea-level rises might require the relocation of 76 million eco-refugees by AD 2030; add to this the impact of altered climate patterns, and the future scenario could be serious – a quarter-century is not long to prepare for these sorts of numbers, and the situation might be far worse if warming accelerates. It was noted earlier that a shift in the Gulf Stream might be caused by global warming, perhaps at about the same time sea level change starts to have effect, and this might turn large parts of Europe and Scandinavia into harsh environments. Large volcanic disasters (also nuclear contamination accidents or warfare) could generate eco-refugees at any moment. Even slight sea-level rises could put some nuclear power stations at risk; tsunami damage is already a real threat. Even a relatively limited natural disaster like the 1988 Armenian earthquake led to 0.4 million relocatees. Global warming may have beneficial impacts, but much depends on the scale and the pace of change.

7.3 Improved awareness and responses to environmental changes

Between the mid-1970s and late 1980s the West became far more aware of environmental issues; Glantz (1994: xi) described the 1980s as the 'beginning of an Age of Environmental Enlightenment'. The reasons for this attitude change are difficult to unravel, but probably reflect improvements in media coverage, increasingly obvious pollution and other eco-disasters, and awareness – as a consequence of space exploration – that the Earth is finite and vulnerable; all this being stirred by the development of environmental activism (**environmentalism**). Environmental concern was developing by the 1920s, but the Economic Depression, the Second World War and Korean War delayed things and the Cold War (*c.* 1947–1993) slowed progress. Between the 1972 UN (Stockholm) Conference on the Human Environment and the 1992 UN (Rio) Conference on Environment and Development there was a huge sea-change of interest in environmental matters. At the start of the twenty-first century there are bodies in place that offer some hope that environmental management can be improved.

Sustainable development has become one of the main goals of environmental management. The literature generally argues that there are four key elements needed to achieve sustainable development:

1. build and maintain human capital – invest in education, skills, health, knowledge
2. build and maintain social capital – maintaining useful traditions, institutions, networks, obligations and cooperation
3. build and maintain economic capital
4. build and maintain environmental capital – sustain the quality of air, water, soils, biodiversity, etc.

In practice, each local strategy will have to be 'evolved' to fit particular needs, threats and opportunities; they will need to be coordinated to ensure they do not conflict with strategies elsewhere, and if possible to try to complement and support them. Much of the work on sustainable development has focused on ensuring that livelihoods continue without breakdown, and upon making economics and energy policies less environmentally damaging (de Haan, 2000). Less attention has been paid to ensuring that sustainable strategies are 'proofed' against natural challenges and make best use of environmental opportunities. Even if a sustainable development strategy is achieved it will have to be subject to careful, forward-looking management in order to survive challenges, some of which could well be natural problems. Efforts should be made to ensure that crucial inputs and services (gene banks, libraries, vital raw materials and skills) are duplicated, dispersed and, if necessary, kept in secure and hardened sites, so that failed strategies can be repaired or replaced.

Key reading

Glantz, M.H. (ed.) (1994) *Drought Follows the Plough: cultivating marginal areas.* Cambridge University Press, Cambridge.

Harding, J. (2000) *The Uninvited: refugees at the rich man's gate*. Profile and London Review of Books, London.

Ulijaszek, S.J. and Huss-Ashmore, R.A. (eds) (1997) *Human Adaptability: past, present, and future*. Oxford University Press, Oxford.

|8|

The future

Nature to be commanded must be obeyed.

Francis Bacon *Novum Organicum* (AD 1620)

Chapter summary

Having explored the interrelations between the environment and humans, are there things that can be done to reduce people's vulnerability and to improve their quality of life? Study of the past suggests possible threats; however, hope of wholly reliable prediction are unlikely to be satisfied – the future is uncertain. The best way forward is vigilance, willingness to at least make contingency plans and, perhaps most crucial, considerable effort to improve human adaptability. In the future nations should focus on development for vulnerability reduction and improved adaptability.

At the start of the last decade of the twentieth century, a group of concerned historical ecologists published 'The Santa Fe Accord' (13–19 October 1990). A key passage runs as follows (Crumley, 1994: 245):

> We are alarmed at the current peril to humanity and the biosphere. ... Historical understanding is essential to enable present and future generations to live with dignity on a changing Earth ... we call for research and management options that draw lessons from the past to guide and provide sustainable quality of life.

8.1 Trying to predict likely future scenarios

Before there can be much useful prediction of the future there is a need to establish 'the state of the world' today, and acquire better knowledge of likely ongoing hazards and opportunities by understanding the Earth's structure and function, the trend of human demands, and through study of the past. Certainly, knowledge of past environmental change has greatly expanded in the last few decades; there is growing knowledge of how the environment functions – but it is woefully incomplete. Predicting future scenarios demands an awareness, not only of possible environmental hazards and trends, but also of likely social, economic, political and technological – in short all – developments (Cooper and Layard, 2001). It is a far from accurate art; but, if there is to be any hope of precautionary environmental management, secure sustainable development and adequate disaster preparation, its development is crucial. Unfortunately, much of the work on future scenario prediction has been based on inadequate research and less than objective assessment; consequently the field has a tarnished reputation. The approach to the millennium saw attempts to predict future scenarios by various authorities (Dooge *et al.*, 1992; Kennedy, 1993); the following are quite typical.

- Humans were pressing the Earth to the limits of its capacity and break-down will occur.
- China and India were rapidly industrializing and had fast-growing populations. What happened in those countries would have worldwide impact.
- With its population expanding, food production in decline, and environmental degradation getting worse, Africa stands out as a likely disaster continent (Parry *et al.*, 1999).
- Japan is the world's largest donor of foreign aid and is likely to influence what is spent on environmental issues in developing countries.
- Large companies and rich countries will get access to the biological resources and biotechnology needed to respond to developmental and environmental challenges.
- In some respects, developed countries with rising labour costs and ageing populations may not be as adaptive as they were in the past.
- There is a shift of economic power to the Pacific Rim; consequently peoples with a non-western, non-Christian (possibly nominally Confucian) viewpoint will have more say in world development.
- Humans have been causing huge damage to soil (since *c.* 1950, roughly one-fifth of the world's cropland has been lost) and over the last 50 years about one-fifth of the tropical rainforest has been cleared; as a result, a vast quantity of biodiversity has been lost worldwide. This is endangering adaptability.
- Global climate change may have marked impact.

There are regular attempts to predict likely future scenarios by bodies like the WorldWatch Institute (see www.worldwatch.org/foreign/index.html). The task is getting no easier – in the past there was only the need to worry about natural challenges; increasingly there is a need to control nature because it has been disrupted by what modern humans call development. Since the 1960s many environmentalists have adopted an apocalyptic tone, warning of impending disaster. They have in the main stopped at whistle-blowing and advocacy; practical monitoring and control measures need urgent attention. The turn of the new millennium prompted a good deal of stock-taking and future scenario prediction (Chasek, 2002; van Ginkel *et al.*, 2001). However, although there is too little practical planning and management, and disagreement between cornucopian (optimistic) and catastrophist (pessimist) environmentalists, there is consensus that there is limited time to act. In 1990 the WorldWatch Institute gave 40 years for the transition to be made to an 'environmentally stable society'; various timespans are suggested by others, but few are more than two generations from the present. The present decade probably offers the last chance for humanity to control its environmental damage and improve how it adapts to natural challenge (Box 8.1). Some hope the transition will come easily. One such optimist was Lomborg (2001), but his predictions have attracted strong criticism from many environmentalists and economists.

Many natural scientists assume that in time it will be possible to model natural processes and get reasonably reliable predictions. That is probably wishful thinking; as Eisenberg (1998) noted, 'scientists and artists often fall in love with their models'. How people and economies respond to natural change is difficult to predict. Recently worries have been expressed in the UK that insurance companies can no longer rely on traditional actuarial analyses because patterns of natural disasters are changing as global warming takes place. So, faced with the risk of 'unexpected' severe natural

---

Box 8.1 Population versus environmental limits

Estimates of the Earth's **carrying capacity** (see Glossary) must be treated with caution because they are forecasts based on interpolating the trends of a huge complex of population, environment, economy and cultural factors. Also, carrying capacity depends on the standard of living people seek – many can live a life of hardship, few a life of luxury. A population that is not viable in the long term might be sustained for long enough for a relatively painless transition to be made to lower numbers – a number of estimates speculate that an 'overdraft' or overshoot situation is likely. A natural or human disaster can upset things and make a 'sustainable' carrying capacity unviable – flexibility must be built in.

Various attempts have been made to calculate global human carrying capacity using data on nutritional and other human needs, and the Earth's productive capacity; Chambers *et al.* (2000: 50–1) plotted 63 selected estimates against historical and projected human population growth. Their conclusion was that a reasonable estimate lies somewhere between 7700 million to 11,200 million. The limits appear to be very close. Given the dismaying rate of biodiversity loss and growing environmental degradation, the limits for maintaining anything like present environmental conditions and allowing people a good lifestyle may well be much lower, below present population – perhaps as low as 1000 million.

disasters, arrangements were made for predictive modelling of climatic extremes and other natural phenomena by a group (known as TSUNAMI) composed of universities, research institutes and the Meteorological Office (*The Times*, UK, 16 September 1996: 14). The international community is going to have to make more similar arrangements for pooling its resources.

It is possible for social and economic change to be intended as well as unintended, so to some extent the future is 'engineered' today (Berhout and Hertin, 2000). As the twenty-first century unfolds there might be increased human understanding of hazards, better technology and new strategies for adaption. But there is also a growing world population, little likelihood of an end to warfare and civil unrest, worsening environmental degradation, and less obvious hindrances like commercialization and public awareness leading to demands from and challenges to policy-makers, but not necessarily willingness to act in sensible ways. The problem is to adapt to rapid and unexpected developments in a world that is not very adaptable.

Woolfson (2000: 192–3) argued that humanity's evolutionary success was in large part because of its ability to reflect, think ahead and choose a course of action in advance; he noted that the bulk of the world's population ascribe to the 'western world-view', which he thinks is no longer appropriate. So, like many other present-day gurus, he called for a shift in thinking to a new worldview. The world, he believed, is at a crossroads: either a new worldview 'takes off' or humans suffer and decline. Part of this new worldview is support for sustainable development and a more 'Gaian' approach that encourages caution and the consideration of wider-scale and longer-term consequences of human actions.

Although futures studies and forecasting are often seen as speculative there has been growing interest from many disciplines, and this has spawned journals such as *Futures*, *Long-Range Planning and Forecasting*, *Foresight*, *Futuribles* and *Global Environmental Change*. Academics and policy-makers have also started to explore strategic environmental assessment and other forms of forward-looking impact assessment applied at the national or international policy or programme levels, rather than at the local, project level. This could help develop a systematic, forward-looking and wide-ranging approach to planning and management.

Identifying likely future scenarios is a more realistic goal than seeking detailed and accurate predictions, or forecasts and tools for doing it: the Delphi technique (a way of seeking consensus from a diverse panel of experts in order to discern likely future scenarios) and strategic impact assessment, for instance, are already well developed (Barrow, 1997). There has already been considerable attention directed to establishing likely global warming scenarios. During much of the post-1945 period governments have been more worried about the ability of humans to exterminate themselves with modern weapons than about the risks of environmental change or natural disasters. There is a need to worry about both. Indeed, natural disasters or environmental change could trigger competition for resources, generate eco-refugees and trigger conflict. Many of our present livelihood strategies are not easily adaptable if environmental conditions alter.

The following 'likely' future scenarios are widely identified (Rivers, 1988: 200; Meadows *et al.*, 1992); however, rather than just one of these materializing it is probable that all will appear in various places.

- *Business as usual* – no dramatic change in human ways or the problems faced.
- *Disaster* – catastrophic environmental and socio-economic breakdown, with a marked decline in human wellbeing and numbers.
- *Ecofascism* – disaster averted by totalitarian controls, and authoritarian rule.
- *High-tech miracle* – new technology overcomes problems.
- *Sane, humane environmental solutions* – humans alter their ethics, develop socially and in a less materially wasteful way; skilful environmental management and sustainable development. The dream of most environmentalists.

Alas, humankind is most unlikely to shift quickly to more environmentally friendly ways; the 'business as usual' scenario seems more likely. Lomborg is not the only optimist, a late 1990s Rocky Mountain Institute (USA) study concluded that the world could industrialize to West German levels with 8000 million people, enjoying a gross product growth of about fivefold, and *still* use one-third less energy than was consumed in 1998, by adopting efficient technology. If such predictions are right human impact on nature might not go out of control.

Ice age or greenhouse world?

Since roughly 2 million years BP the Earth has undergone glacials of about 100,000 years' duration, separated by warm interglacials; if the natural pattern seen in the past continued there would probably be a return to ice age conditions within approximately 1000 years. That is now unlikely because there has been anthropogenic global warming. With carbon dioxide levels presently approaching 400 ppm, the naturally 'scheduled' ice age will probably not return and in the long term moderately warmer conditions may benefit the world (granted this is speculation).

There is a possibility of serious global warming leading to considerable sea-level rise and disruption of food production. Present gentle warming might start a positive feedback of some sort, say sudden outgassing of methane from marine sediments, which magnifies the trend (and perhaps speeds the rate of change), even leading to positive feedbacks and perhaps a 'runaway' situation. Nature supported greenhouse conditions in the past, so there is always a risk that anthropogenic changes might trigger similar conditions. Roughly 55 million years ago, global temperatures were probably about 15°C above present. Humans and many other organisms can adapt to considerable change, but it takes time if there are to be no catastrophic effects; marked temperature change over anything less than several decades would be problematic, as would regional cooling triggered by global warming.

Until quite recently fears were mainly about a return to cool Little Ice Age-like or even full ice age-like conditions. In 1973 a group of eminent scientists wrote to the US President warning him about the risk of an ice age developing within a few centuries; the same warning was still being voiced in the 1980s (e.g. by Hoyle, 1981a; 1981b) and is occasionally heard today. Palaeoecological evidence indicates that some past climate changes, including cold phases (stadials), started and finished suddenly. Whether full glacials started suddenly is uncertain; there have been reports of glacial deposits in southern England that suggest it did, including claims for signs of glaciers in Somerset and the English Channel, which flowed west–east, which may hint at ice age origins in severe snowstorms from the west. In western UK, fossil beetle remains indicate that very sudden and marked climate changes have taken place in the last 10,000 years; these shifts taking just a few years. It may be that glacials begin in uplands with ice caps accumulated *in situ* as snowfalls fail to melt in summer. More sensationalist writers have termed this 'ice blitz': suddenly a colder than average winter brings snow, which fails to melt the following summer; this is repeated, and by the time the population of western Europe or the higher latitudes in general realize, it is too late – they would have to become eco-refugees or perish (and eco-refugees when, perhaps, transport is disrupted and conditions hazardous). Whether lower-latitude countries would accept a sudden influx of huge numbers of relocatees in what would probably be a period of climatic disruption for them as well, and a time of economic and agricultural chaos, is far from certain. It might be that people would delay evacuation too long and be too debilitated for more than a few to escape.

Nowadays, fears about climate have shifted to the opposite: global warming. The theory of greenhouse warming was discussed in the nineteenth century, but not seen as a real possibility until it was realized that atmospheric carbon was increasing (G.S. Callendar had voiced such concern by 1938). One environmentalist noted the not uncommon sensationalist headline of the last decade – 'global warming – Earth's climate out of control' – and observed that people forget the world's climate has never been under control! Worry about global warming is acceptable – the public and policymakers should take it seriously – but there is less awareness that there is a chance that sudden natural cooling could still occur, and might even be triggered by global warming. While the Earth as a whole may benefit from global warming – in the long run – there might be a sudden weakening of Atlantic circulation that cuts the flow of the Gulf Stream to Europe (see Chapter 5). Western Europe and possibly eastern USA could then face a disaster. The question is, will ongoing global warming counteract natural cooling and be beneficial, or will it trigger difficulties like sudden cooling of western Europe and eastern North America? In the longer term, especially if there is no perturbation of the Gulf Stream, global warming to pre-Quaternary levels would be vastly preferable to a return of glacial conditions, which nature might have delivered were it not for human activity. However, the change en route to such conditions could be catastrophic.

Even quite minor change is difficult to adapt to if it happens fast, whether it is warming or cooling. The governments of developed countries appear to be assuming that the future is one of relatively gradual warming. There is no guarantee that change will be gradual and easily manageable; the past suggests it is possible that it could be sudden, and it might be cooling, not warming. In a recent issue of *The Times* (UK, 7 September 2001: 8) Mark Henderson observed that 'The planning needed for rapid cooling is very different' to that needed for dealing with gradual warming. *The future is uncertain*; we cannot be sure what will happen, and most countries today are poorly prepared to adapt.

With an uncertain future there should be more expenditure to improve food production in as many different parts of the world as possible, and efforts to stockpile much more. That would give tangible benefits immediately: more food surpluses in store now to buffer famines and allow food aid, whatever the cause of shortage, and to protect against occasional disasters like mega-volcanic eruptions or extraterrestrial body strikes. Stockpiled food supplies would also give some hope of adapting to sudden unfavourable climatic change. Struggling for greenhouse emission controls, tradable emissions quotas and so on might be well intended, but is probably ineffective expenditure of effort and a great deal of resources. In practice it may be blighted by difficulties over agreements and enforcement (and such spending does nothing to protect the world from occasional natural catastrophes). Better to concentrate on predicting what changes will take place and improving chances of adaptation to environmental change; global warming is already initiated, it makes little sense trying, at huge cost, only to prevent or reduce emissions, yet doing little to adapt to change already underway and to largely ignore other threats. *If it is uncertain what to prepare for, then the priorities should be to improve adaptability, especially by increasing emergency food reserves.* A priority should be to boost resilient and sustainable agriculture in as many places as possible so that more food can be produced and saved. There is at present little incentive for authorities and food producers to produce more and stockpile it. When the EEC, largely unintentionally, accumulated stored surpluses ('food mountains') a few decades ago, it generated considerable media criticism and economic impacts when it was 'dumped' on the market. How can such goals be achieved in a world where commercial cropping for export gets funding but smallholder sustainable agriculture gets little?

Global warming is likely to raise sea-levels – what is not yet clear is by how much and how fast. There have been a number of projections, one by

Milliman *et al.* (1989) suggested that Bangladesh and the Nile Delta lands face as much as 4.0 metres' rise by AD 2100, which could mean displacement of roughly 35 per cent and 24 per cent of the populations of these countries (respectively) – many millions. More recent estimates seem to confirm these estimates. Given that, with current sea-levels, cyclonic storms and river flooding can cause terrible impacts, almost any rise is unwelcome. Large sums of money have been spent trying to construct flood defences for Bangladesh, many would say with limited success, and that it might be better to concentrate more on cyclone and flood refuge construction and warning systems. Whatever is done, there is a real risk of large numbers of eco-refugees who will need relocation. Can the international community prepare suitable strategies for relief and, if need be, resettlement for vastly greater numbers than have been dealt with so far?

Although global warming has not been proven to have directly killed anyone, billions of dollars are still being spent on it. Other real threats are virtually ignored: expenditure on soil erosion control has been cut in a number of countries in recent years, and pollution of the global environment deserve much more funding. The control of land degradation, improvement of land husbandry, and conservation of soil and water should be given high priority.

8.2 Better preparation for environmental change

One of my aims in writing this book was to highlight the need for more attention to be paid to natural hazards and opportunities. There is another pressing need, and one that is difficult to separate from natural change: human-induced (anthropogenic) change. Focusing on this, Stone (1993: 3) argued, perhaps a little overdramatically, that 'The Earth has cancer, and that cancer is Man.' That point has been made many times before, but Stone went further and argued that the post-1945 Green Revolution efforts to boost agriculture could be seen as supplying the 'growth phase of the human cancer of Earth'. He also noted that after growth some malignant cancer tumours tend to undergo local necrosis, and that could be paralleled in what is happening to many livelihood strategies, degradation of inner cities, ruination of ocean fisheries, etc. The colonial era, Stone suggested, can be seen as the metastasis phase, during which western attitudes (the cancer) spread globally to become established at many new sites. Will the 'cure', reaching a balance with nature and practising sustainable and adaptable development, be possible with gentle democratic measures or does it require drastic 'surgical' treatment?

Climate change has attracted interest (alas other fields have had less). In the UK, and elsewhere, multidisciplinary centres have been founded to try to establish the future implications of climate change (in the UK these include the Tyndall Centre for Climate Change at the University of East Anglia, and the Environmental Change Institute, Oxford University). The hope is that it will be possible to identify periodic and quasi-periodic threats, thresholds that reward monitoring, and other signals, all of which offer some early warning. Recently the International Panel on Climate Change (IPCC) has been trying to assess future climate change impacts and likely adaptation (Parry and Carter, 1998). A number of others have also explored future climate scenarios (e.g. Ruddiman, 2001). But, even if research progresses well, there are always going to be surprises.

The 'precautionary principle' is central to modern environmental management, and is promoted as a common-sense approach that reduces mistakes and delays in responding to problems. If it is to be effectively applied then decision-makers should not wait until there is clear, unrefutable proof that a problem has developed, for by that point it may be too late to act. Therein lies a decision-making dilemma: to act too soon means it can be difficult to carry support and if the fear proves to be a 'false alarm' the decision-maker suffers loss of credibility (and possibly dismissal); delaying until there is proof means the problem is more costly to resolve or is incurable. Even the most adept decision theorists find this quandary very difficult to resolve. Eisenberg (1998: 299) concluded, after a fascinating evaluation of environment–human attitudes, that in the future environmental management must be flexible and responsive and 'ride the waves of nature's changes'.

Faced with uncertainty, some human activities need to be undertaken with great caution; for example, nuclear waste repositories should be constructed where they will not be breached by even the most severe earth movements, well above the reach of even the most unlikely mega-tsunami, safe from terrorists and accidents, and where they will escape events like ice ages. They must also offer storage or containment even without human supervision. Crucial resources like gene banks, agricultural seed stores and libraries should be duplicated and kept in safe locations so that a major natural disaster would not destroy everything.

There are a great many different potential future problems, necessitating multidisciplinary 'wide spectrum' monitoring and forecasting if there is going to be any avoidance or timely response. There may be some potential in regional risk assessment or regional impact assessment approaches (Walker *et al.*, 2001). The hazards the world faces are real enough, but there is so much uncertainty over likely location and timing that preparations will be difficult to promote. With many possible hazards the crucial threat is disruption of food production for one or a few years. As already stressed, a wise precaution would be to stockpile much greater amounts of emergency food supplies than are currently held.

The threat of future disease epidemic is significant, and one against which more resources have been directed, partly because a number of governments fear biological warfare or terrorism. Predictions that global warming will cause marked changes in disease patterns are still largely speculative; Reiter

(2000a; 2000b) urged caution, arguing that the shifts in disease patterns seen so far are as likely to reflect better surveillance and reporting. He also noted that malaria in Europe continued to be transmitted in the Little Ice Age and that decline of transmission seems to have been through rural depopulation, land drainage, better building standards and cropping changes (at a time when temperatures had risen). Consideration of environmental issues must be included more effectively in planning and policy-making, but simplistic predictions must be treated with caution.

8.3 World food security

Developed country peoples tend to see food surpluses and food aid as charity for others – they (the rich) are in control, the North helping the South. It might well be that the time will come when the South will be asked to help the North: if, say, there were a sudden shift of the Gulf Stream. By investing much more than at present in improving lower-latitude agriculture and development in general, developed countries would boost their own survival chances as well as helping today's poor in developing countries. A well-fed Third World would be less likely to suffer epidemics or unrest that could spread to affect the developed countries – immediate benefits for the donors. Developing countries that were self-sufficient in food would better ride out the chaos of a global disaster, and might be able to help developed countries (most of which are at a higher latitude where they may be more vulnerable to cooling).

Today the world relies on grain surpluses from just a few regions, some of which suffer frequent rainfall fluctuations and ongoing soil degradation, and the rest might be disrupted by even minor climate change (wheat production in the CIS and North America is vulnerable to climatic change). Unfortunately, developed countries seem unconcerned about food supplies. Yet, as recently as 1972 and 1975, the USSR had to make huge purchases of grain on the world market because its production was hit by drought, land degradation and poor planning, and that sent world food prices soaring. At roughly the same time there was a major food supply problem in China. The North American prairies, one of the world's major food-producing regions have been developed during a period of relatively favourable climate; the assumption, there and elsewhere, has been that modern seeds and agricultural technology can match any likely climatic deterioration – that is dangerous. Also, there is too little concern for ongoing soil degradation and the problems caused by dependence on agrochemicals to maintain yields.

There are already many studies of the likely effects of global warming (for a recent review see Harvey, 2000a; 2000b). Warming would probably mean drought hitting grain production in the wheatlands of North America, Russia and Western Australia, and even if global warming allowed crops to be grown further north, there might not be suitable soils. Breeding new varieties of crops suitable for the altered conditions may take decades so there could be problems adapting to rapid environmental change (it would be wise to try to predict likely changes and find ways to speed up responses). The impact of raised carbon dioxide levels on crop transpiration (and hence moisture demands), on photosynthesis (affecting overall growth and ratio of straw to grain), and on competition between weeds and crops is not clear. Phytotron experiments (i.e. controlled-environment chambers) and modelling suggest that some crops may benefit from moderate carbon dioxide increase, while others will suffer. An unwelcome response to raised carbon dioxide levels and warming might be more vigorous weed growth and depressed crop photosynthesis. Crop pests like aphids and fungal infections are likely to be problems if there is global warming.

More secure food production is essential and should be a priority. There is no shortage of calls for improved and sustainable food production, but less attention to safeguarding supplies against disasters.

Lutz (1996: 3) predicted that the world's human population will reach 6500 million by AD 2006; there have been estimates of the Earth's ultimate carrying capacity, ranging from 6000 million to 80,000 million. Many worry that we are already having trouble sustaining food production and that demographic growth will set a difficult, if not impossible, target; therefore any hope of producing larger surpluses for storage as a safeguard against future emergencies could be wishful thinking. Projections suggest that yields might increase through global warming at high and mid-latitudes, but fall at lower latitudes.

A number of commentators have voiced concern that growing urban demand, pollution and global warming will mean greater competition for irrigation water, so that there will be a slowing of any further agricultural expansion and possibly a fall in production. Nevertheless, successes in countries like Israel and Finland have demonstrated that there can be productive agriculture in harsh environments. And, in developing countries like Kenya (notably Machakos District), farmers have been shifting to keeping livestock penned; the accumulated dung can then be spread on cropland, boosting yields, helping to halt soil degradation and improving security against drought (because of the better moisture-holding capacity of the improved soil). This is hardly a technological breakthrough, but it has proved very effective, is spreading fast with little outside interference and investment, and gives some hope of improving yields enough to mean less dependence on international grain surpluses (which may allow some storage for security against disasters). There are also signs of hope in Zambia, where both larger and smaller farmers have recently been developing organic farming, using compost to maintain production and improve the soil. With enough improvement of farm extension services and some limited support, the trend may take off and lead to self-sufficiency, sustainable production and surpluses against disasters. One can only hope that these improvements are the first of many. (For further information on soil fertility management in Africa see Scoones, 2001, especially pages 176–214.)

There are other innovations that may have potential for easing future food shortages and making storage of more emergency supplies a possibility, even in poorer countries. One of these is the spread in use of green manures (vegetation planted to improve soil directly or indirectly through composting), which largely eliminate the need for chemical fertilizers, cutting pollution, improving the soil and making farmers less dependent on costly inputs. A promising green manure is the mucuna bean (*Mucuna pruriens*). In parts of Central and South America there have been encouraging successes in sustainable food production and land rehabilitation through the adoption of this plant into the cropping systems of both small-scale farmers and large commercial producers. There are signs that the approach is spreading without much official intervention. Faced with uncertainties over many aspects of food production, the promotion of innovations like mucuna – which appears to enhance the soil effectively, make farmers less dependent on chemical inputs, and promises more secure cropping and increased food production – deserves much more support.

Although not a new concept, Charles Paton has pioneered workable cost-effective designs for low-energy-need greenhouses, which are chilled by cold sea water to catch and condense enough atmospheric moisture to irrigate crops. These are suitable for water-scarce coastlands and islands where they could produce food even though there is little or no available irrigation water. Given the pressure on the world's water supplies this technology has great value, if the funding can be provided to start it up (Pearce, 2002).

There have been attempts to predict future food production, but the world food system is complex and dynamic, with physical, technical innovation, economic, political and social factors interacting; reliable prediction is impossible (Döös and Shaw, 1999). Crucially, the world food system is geared to meeting current market demand (commercial) for food, not to looking ahead and stockpiling in case of disruption. To make much progress toward the latter goal will almost certainly demand deliberate interventions to encourage extra production and to support the costs of transport and storage.

A strategy that might support the intensification of agriculture, and one that would help avoid exacerbating the urban–rural divide that is growing in many cultures, is community-sponsored agriculture (CSA). CSA consists of a group (anywhere from a score to hundreds) of urban, or at least non-farming, families cooperating and sharing the expenses of a farm or farms and guaranteeing the farmer(s) a basic income. The farmer gains some security, and the supporters get control over the quality of their food and also access to recreational use of the land – this can include help with some of the farming tasks, barbecues, etc. The end result can be much more intensive agriculture with little state outlay, recreational benefits for the community, and improved awareness of and support for rural activities.

Eisenberg (1998: 379–83) tried to identify ways forward to achieving more sustainable and secure living conditions, and reducing environmental damage. Two broad options have been suggested.

1. *Decoupling* – the world's environmental managers should seek to concentrate human populations as far as possible in cities to minimize urban sprawl and the loss of productive land beneath buildings. Advanced technology would be used to cut pollution and reduce demands made upon nature. Hopefully, adequate areas of the world could then be left alone to revert back to 'wilderness'. There are risks in such a strategy: urban dwellers (as is already the case) would lose touch with nature, and governments might have difficulty making them pay enough for food and other resources, and for nature conservation. It may also prove difficult to restrict people to city areas sufficiently.

2. *Return to Eden* – the approach favoured by many radical environmentalists. Discard technology and pursue a simple lifestyle. With a world population of around 6500 million, any such strategy would spell disaster, leading to widespread loss of life and probably environmental damage during the painful transition (Lewis, 1992). The danger is that misguided or misdirected popular opposition to technology and 'top-down' environmental management could become powerful. Such developments would probably ruin the chances of halting environmental degradation, and of improving human wellbeing and survival opportunities.

Development for vulnerability reduction

Modern 'progress' has generally meant the growth of high-density population centres, supported by agriculture reliant on increasingly intricate technology and trade, mainly under the control of elites, generating pollution and failing to sustain the soil (Crumley, 1994: 185). In the face of environmental uncertainty, modern agriculture should pursue flexibility, adaptability and resilience, as well as seeking sustainable production and a reduction of damage to the environment. Most efforts have been directed at agricultural yield increase; more recently some attention has been given to sustainability, but adaptability, diversification and security have had much less support.

Today humans dominate the Earth and affect much of its functioning: we tap much of the primary productivity, affect the climate, have damaged the stratospheric ozone layer, cut forests, pollute the environment and decimate biodiversity. Unfortunately, human dominance is not control: 'blind humanity steers the globe'. And if we do not improve stewardship and fit in with what we have left of the biosphere it will probably collapse 'and excrete us from the wreckage' (Nisbet, 1991: 299). The challenge is to manage resources and environment–human relations under conditions of uncertainty (Holling, 1978; Walters, 1986).

The speed of change (natural and human-induced) and the uncertainties make established approaches to problem-solving more or less obsolete (Simmons, 1997: 41). The globe is increasingly interdependent, pollution crosses borders, most localities depend on others and their actions interact. Any attempt at management is faced with complexity, and will have to address biophysical, socio-economic, political and cultural issues. There will need to be a quick and effective response to challenges in spite of much uncertainty. Managers will have to simultaneously consider all sides of problems: local, regional, global and different processes proceeding at varying speeds (Simmons, 1997: 41).

Botkin (1990: 193) warned that the prevailing view of nature is mechanistic, making the assumption that the environment is something that can, if the need arises, be 'repaired'. That, he argues, is a recipe for disaster – to survive, humans must better understand nature, 'fit in better' and be prepared to adapt fast (see Box 8.2). He also asked how humans can effectively manage nature, which is complex and always changing. The

---

Box 8.2 An assessment of failure to adapt to environmental and economic changes

Drawing on a study of the misfortunes of the Norse settlers in Greenland (see Chapter 3), McGovern (in Crumley, 1994: 148–9) tried to assess why there was a failure to adapt to environmental and economic changes. Administrators contemplating the future and the need for human adaption should be aware of these points. *Note:* Suggested remedies are in {parentheses}.

- *False analogy* – the problems in Greenland were judged against Icelandic or Norwegian experience and missed critical differences. {Question the relevance of hindsight.}
- *Insufficient detail* – managers overgeneralized. Resources were patchy, not as abundant as imagined, nor as resilient as hoped. {Question data.}
- *Short observation time* – the Norse decision-makers had too short a memory (plus records) to track significant environmental variations. {Make use of palaeoecology and environmental history and be willing to spend on long-term monitoring.}
- *Managerial detachment* – decisions were made by an elite that was spatially and socially separated from the public in general (especially producers). True of the increasingly urban and developed country focus today. {Increase accountability of management and empower people.}
- *Reactions are out of phase with threats* – mainly because of the previous two points. {More precautionary planning.}
- *The problem faced is 'someone else's problem'* – managers do not feel threatened so do not act; indeed, difficulties might even weaken their rivals and troublesome subordinates to the managers' advantage. {'Police' the management.}

Source: Author's interpretation based in part on the views of McGovern in Crumley, 1994.

likelihood is that, at present, nineteenth-century perceptions and models are being applied to twenty-first-century problems (Botkin, 1990: 6).

It is only recently that enough has been learnt about the Earth's structure and function to make effective environmental management a realistic goal; before the 1957/58 International Geophysical Year there had been limited exchange of information between nations. In the last few decades enough common measurement units have been adopted and sufficient environmental parameters have been internationally defined to make global environmental monitoring possible. That goal has been helped further by the end of the Cold War, improved computing, better and cheaper telecommunications, satellite remote sensing, and a will for peoples to share data and jointly manage environmental problems. Forward-looking modelling of environmental change is being undertaken using closed-chamber experiments, the largest and most controversial being the Biosphere 2 facility in Arizona (USA). After much disagreement and many management changes, Biosphere 2 has shifted from sensationalist experiments of limited practical environmental management value to more rigorous study of global ecological issues.

History strongly suggests that predicting a threat may not be enough to get decision-makers to act and tax-payers to pay to support avoidance or mitigation. Policy-making and the attitudes of citizenry have to evolve to react to threats that are not 100 per cent proven. There is a need to increase support for 'civil defence' but not, as between the 1940s and 1990s, focusing mainly on mitigating the effects of warfare. What is needed is better preparation for natural disasters. The approach could be similar to that of the Internet: a set of diverse independent preparations that are coordinated globally.

Recently there has been growing interest in large-scale hazards; a point made by Fagan (1990) and others is that while we should certainly try to prepare for such 'mega-threats', it is also vital to treat more mundane 'everyday threats' more seriously than is presently the case. Soil degradation is relatively neglected, yet we largely feed on produce that is often grown on rapidly degrading land. Technological progress must help human adaptation, blind neo-Luddite rejection of biotechnology and other modern methods will not feed the hungry, nor protect the environment. Use of technology must be what Lewis (1992: 147) calls 'Promethean' – that is, it is used boldly, but with great caution and environmentally sensitive vision. Moss (2001) reached a similar conclusion, that humans must use biotechnology to manage the Earth's environment, not reject the opportunity and hope for the impossible: that left alone the world will heal itself.

The challenge is how to improve environment–human interrelations, to gain better environmental stewardship, and improve human quality of life and chances of long-term survival (Fig. 8.1). History has yielded up examples of situations where, at island or regional scale, people have failed to adequately meet this challenge (e.g. the overpopulation and degradation of Easter Island during the last millennium, or the medieval Greenlanders

---

(a) **Past** (pre-human, early humans) ARCADIA

Earth/biota functions naturally.
Occasional terrestrial and extraterrestrial perturbations, some posing a serious threat to organisms.

(b) **Present** – DYSTOPIA

Earth/biota much altered by humans.
Humans are not in control and have caused breakdown or distortion of many natural functions and environmental controls (crisis situation?). Environmental degradation widespread and ongoing loss of biota. Occasional terrestrial and extraterrestrial perturbations, some posing a serious threat to organisms.

(c) **Future** – UTOPIA

Earth/biota well managed. Degradation halted (ideally reversed), conservation, stewardship.
Occasional terrestrial and extraterrestrial perturbations, some posing a serious threat to organisms. Humans increasingly able to intervene and counter threats, or at least improve chances of surviving them.

Figure 8.1 Environment–human interrelations: past, present and (possible) future
Source: Author, using material from various sources.

apparent failure to respond satisfactorily to climatic and economic setbacks). There is a need to treat natural threats more seriously, to try and make humans at least as adaptable as they were in the past, and to move to adequate environmental stewardship. This means skilfully managing environmental and social problems, and those skills are far from well developed. Methods of assessing how humans impact upon the environment are improving – one promising field is **ecological footprinting** (see Glossary) (Chambers *et al.*, 2000). It would be useful if planners could assess social capital and impacts likely to affect it, and monitor people's vulnerability to natural threats. Amongst the advocacy some routes seem to be opening; one may be to adopt 'political ecology' approaches (Kiel *et al.*, 1998: 29; Stott and Sullivan, 2000). Political ecology seeks to analyse problems and opportunities, find solutions, and manage the human and environmental factors (using technology where appropriate).

A potential strategy is for environmental management to be practised on a relatively small scale, which should mean diversity and flexibility, more manageable planning, monitoring and governance. Efforts on a nation-state scale are often dismissed as too big to deal with small problems, and too small to cope with larger challenges. The best scale is probably similar to that adopted by the Swiss Federation – canton or county scale. In Switzerland and a number of other countries this seems effective, allowing for a diversity of lifestyles, outlooks and social needs; it also has a 'human scale' (Schumacher, 1973). Ultimately global coordination is vital to ensure that 'cantons' do not come into conflict, to encourage cooperation and duplication of crops and other resources (to enhance adaptability and resilience), to manage research and large-scale monitoring, and other specialist services that are beyond the capacity of the local level. The cantons would seek sustainable development, satisfactory environmental management and, if possible, food and other commodity surpluses. If need be, cantons would be assisted by others – perhaps even twinning-type arrangements – which would have some advantages over some of today's foreign aid. The overall pattern would be similar to a mosaic design (or the cells of an organism), each tile (canton) having a particular form (some differing greatly from others) and all cells making an overall pattern. If one canton gets into difficulties others may resupply seeds, expertise, etc. This strategy is in some ways similar to the ideas published by Sale (1991); however, I would not insist on the units being bioregionally determined (some argue that bioregionalism is flawed by being a crude form of environmental determinism); there will be situation where other criteria should be used. Some cantons might rely on intermediate technology and 'grass roots' inputs, others on the most advanced technology and the services of specialist bodies.

While the way forward is not clear, the goals are, and attention should be directed toward:

- reducing and repairing damage to critical elements of the environment and biota; in particular, loss of biodiversity should be halted as fast as possible to avoid reducing future development options
- putting into place strategies to improve human adaptability and survival if environmental changes happen
- improving environmental stewardship.

Progress is unlikely unless poverty and inequality are much reduced; which is going to be a huge challenge.

Key reading

Cooper, R. and Layard, R. (eds) (2001) *What the Future Holds: insights from social science.* MIT Press, Cambridge (MS).

Eisenberg, E. (1998) *The Ecology of Eden. Humans, Nature and Human Nature.* Knopf, New York (NY), and Macmillan (Picador), London.

Lewis, M.W. (1992) *Green Delusions: an environmental critique of radical environmentalism.* Duke University Press, Durham (NC).

Glossary

AD – 'after Christ', a recent politically correct equivalent is CE (common era). See also BC.

Adaptation – adjustments in ecological, social or economic systems in response to actual or expected environmental change, their effects or impacts. The concept is also used by psychologists, and was central to the (dubious) work of Herbert Spencer (see Chapter 1).

Adaption – this usage is adequate for most purposes: the ability of a population to adjust, biologically and behaviourally, to environmental conditions (Ulijaszek and Huss-Ashmore, 1997: 1).

BC – 'before Christ', the recent politically correct equivalent is BCE (before the common era).

Billion – one thousand million.

BP – before present (present being AD 1950).

Carrying capacity – essentially the maximum number of organisms an environment can support indefinitely. When exceeded there are insufficient natural resources (and with human populations perhaps breakdown of institutional supports).

Climate – this is the 'average' of weather, conditions over the longer term and wider area (and may be seen as characteristic of a particular region). Usually the weather data of the past 30 years are 'averaged' to determine climate. See also *Weather*.

Cultural ecology – study of the adaptation of human societies or populations to their environments, emphasizing the management of techniques, economy and social organization through which culture mediates the experience of the natural world.

Culture – a concept developed by anthropologists to describe the distinctive adaptive system used by humans. There are many definitions, so it is impossible to offer one that is precise and fully acceptable to all; the following will hopefully serve: a society's traditional system of belief and behaviour, understood or adopted by individuals and members of social groups, and manifest in individual and collective behaviour. It is a complex system, a blend of shared traits (tools, customs, skills, beliefs, social organization, etc.) that permits a society to interact with its environment.

Currents – see *Winds*.

Dating – dates cited by various sources may be derived from relative dating techniques like radiocarbon (^{14}C), which may sometimes be statistically 'corrected'; counts of varves or tree rings; and several other dating techniques – so there is often variation from source to source and apparent correlations should be treated with caution.

Dendrochronology – dating, and some indication of past environmental conditions, from tree-growth rings.

Development – a process involving physical, technological and social change. Often (but not always) development is a 'planned' process of change that becomes a framework within which all else (planned and unplanned) happens.

Ecological footprinting – establishing the land and water area that is needed to support indefinitely the material standard of living of a given human population, using prevailing technology.

Ecology – definitions vary, a representative definition would be: the study of the relations between various organisms and their physical surroundings, and upon the adaptions made by the organisms.

Environmentalism – mainly since the mid-1960s there has been a paradigm shift in western countries with growing numbers of intellectuals, activists, professionals and citizenry interested in, and concerned about, environmental matters. Environmentalism is a generic term for a very wide spectrum of attitudes that all share a desire for better stewardship of nature.

Green Revolution – efforts to boost agricultural yields by encouraging and supporting the use of a 'package' of improvements, notably high-yield varieties of crops, fertilizer, pesticide and herbicide. The approach was originally developed by American-funded research and extension stations in Mexico and the Philippines; during the 1960s and 1970s it spread through the developing countries of South and Southeast Asia and Latin America, and later to Africa. The efforts probably prevented famine and kept food production abreast of population increase, but it was often socially and economically damaging to poorer farmers, led to agro-chemical pollution and pest resistance to pesticides, and made developing countries dependent on oil-based inputs and seed companies. Since the 1980s there have been calls for a 'Doubly Green' Revolution, which would not have the unwanted side-effects, especially environmental damage.

Heinrich event – ocean sediment record indicating ice-rafted debris, interpreted to indicate a short-duration (<1000 year) cold (stadial) phase.

Human capital – Based on health, knowledge, skills, motivation, drive, morals, etc.

Malthusian – based on the late eighteenth and early nineteenth-century ideas of the Reverend Thomas Malthus. He suggested that excessive liberalization of Britain's inadequate Poor Relief would lead to population growth that would outstrip food production, leading to disaster. Often interpreted as simply the tendency for population increase to outstrip improvements to food production. Similar views were widely held by 1960s and 1970s neo-Malthusians.

Milankovich – Milutin Milankovich was a Serbian astronomer who, in the 1920s, calculated that the way in which the Earth orbited the Sun explained periodicity in climate changes during the last three million years. The theory has gained much more support in recent years. Similar suggestions were made earlier by James Croll, a Scottish geologist.

Natural hazards – those phenomena in nature that have the potential for causing damage.

NEO – Near Earth Object; a body passing at a distance similar to or less than that at which the Moon orbits. Effectively likely to hit or 'near miss' the Earth.

Pandemic – a disease outbreak affecting more than one continent.

Planetesimal – extraterrestrial bodies, smaller than a planet: comets, asteroids, bolides, meteorites, etc.

Political ecology – this approach has evolved in part out of political economics. The aim in political ecology is to improve environmental intervention and change.

Social capital – the value added to any activity by human relationships and cooperation. Supported through social institutions (families, extended families, communities, cooperatives, associations, businesses, NGOs, schools, trades unions, etc).

Uniformitarianism – the belief that broadly the same internal and external Earth processes that are visible today operated in the past.

Utilitarian – Utilitarianism is a philosophy of life developed in the eighteenth century, especially by Jeremy Bentham (UK), the guiding principle of which is to act to produce the greatest amount of happiness for the greatest number of people.

Weather – essentially local and current atmospheric conditions. See also *Climate*.

Winds – a wind blowing *from* the west is a westerly; from the east, an easterly; from the south, a southerly; from the north, a northerly. However, ocean currents are named according to the direction in which they flow (so a westerly current flows westward, etc.).

References

Aarmelagos, G.J. (1998) The viral superhighway. New York Academy of Sciences (8 pp.), may be downloaded at http://findarticles.com/cf_0/m2379/nl_v38/20136393/print.jhtml, accessed 27 March 2001.

Adams, J., Maslin, M. and Thomas, E. (1999) Sudden climate transitions during the Quaternary. *Progress in Physical Geography* 23(1), 1–36.

Adams, W.M. (1992) *Wasting the Rain: rivers, people and planning in Africa*. Earthscan, London.

Adger, W.N. (2000) Social and ecological resilience: are they related? *Progress in Human Geography* 24(3), 347–64.

Ager, D. (1993) *The New Catastrophism: the importance of the rare event in geological history*. Cambridge University Press, Cambridge.

Agrawal, A. (2001) As sick as a parrot. *New Scientist* 169(2283), 16.

Alatas, S.H. (1977) *The Myth of the Lazy Native: a study of the image of the Malays, Filipinos and Javanese from the 16th to the 20th century and its function in the ideology of colonial capitalism*. Cass, London.

Albey, R. (2000) Warming to a climate of change. *Times Higher Education Supplement*, 6 October, 20–1.

Albritton, C.C. Jr (1989) *Catastrophic Episodes in Earth History*. Chapman and Hall, London.

Alley, R.B. (2000) *The Two Mile Time Machine: ice cores, abrupt climate change and our future*. Princeton University Press, Princeton (NJ).

Alvarez, L.W., Asaro, F., Michel, H.V. and Alvarez, W. (1980) Extraterrestrial cause for the Cretaceous-Tertiary extinction. Experimental results and theoretical interpretation. *Science* 208(4448), 1095–108.

Alvarez, W., Asaro, F., Michel, H.V. and Alvarez, L.W. (1982) Iridium anomaly approximately synchronous with terminal Eocene extinctions. *Science* 216(4548), 886–88.

Alvarez, W., Alvarez, L.W., Asaro, F. and Michel, H.V. (1984) The end of the Cretaceous: sharp boundary or gradual transition? *Science* 223(4641), 1183–6.

Arnold, D. (1996) *The Problem of Nature: environment, culture and European expansion*. Blackwell, Oxford.

Aronson, V. (2000) *The Influenza Pandemic of 1918*. Chelsea House, Broomall (PA).

Arvill, R. (1967) *Man and Environment: crisis and the strategy of choice*. Penguin, Harmondsworth.

Atampugre, N. (1993) *Behind the Lines of Stone: the social impact of a soil and water conservation project in the Sahel*. Oxfam, Oxford.

Athanasiou, T. (1996) *Slow Reckoning: the ecology of a divided planet*. Secker and Warburg, London.

Barrow, C.J. (1991) *Land Degradation: development and breakdown of terrestrial environment*. Cambridge University Press, Cambridge.

Barrow, C.J. (1997) *Environmental and Social Impact Assessment: an introduction*. Arnold, London.

Barrow, C.J. (1999) *Alternative Irrigation: the promise of runoff agriculture*. Earthscan, London.

Barrow, C.J. and Hicham, H. (2000) Two complimentary and integrated land uses of the western High Atlas Mountains, Morocco: the potential for sustainable rural development. *Applied Geography* **22**(4), 369–94.

Barrows, H.H. (1923) Geography as human ecology. *Annals of the Association of American Geographers* **13**(1), 1–14.

Barry, J. (1999) *Environment and Social Theory*. Routledge, London.

Beck, U. (1992) *Risk Society: towards a new modernity* (first published in German in 1989, translated to English by M. Ritter). Sage, London.

Bennett, J.W. (1976a) *The Ecological Transition: cultural anthropology and human adaption*. Pergamon, New York (NY).

Bennett, J.W. (1976b) Anticipation, adaptation, and the concept of culture in anthropology. *Science* **192**(4242), 847–53.

Beresford, M.W. (1954) *The Lost Villages of England*. Lutterworth Press, London.

Beresford, M.W. and Hurst, J.G. (eds) (1971) *Deserted Medieval Villages*. Lutterworth Press, London.

Berhout, F. and Hertin, J. (2000) Socio-economic scenarios for climate impact assessment. *Global Environmental Change* **10**(3), 165–8.

Binford, L.R. (1999) *As the World Warmed: human adaptions across the Pleistocene-Holocene boundary*.

Binzel, R.P. (2000) The Torino Impact Hazard Scale. *Planetary and Space Science* **48**(4), 297–303.

Bischof, J. (2000) *Ice Drift: ocean circulation and climate change*. Springer-Verlag, Heidelberg.

Biswas, A.K. (1979) Climate and economic development. *The Ecologist* **9**(6), 188–96.

Black, R. (1994) Forced migration and environmental change: the impact of refugees on host environments. *Journal of Environmental Management* **42**(4), 261–77.

Blaikie, P.M. and Brookfield, H. (eds) (1987) *Land Degradation and Society*. Methuen, London.

Blaikie, P.M., Cannon, T., Davis, I. and Wisner, B. (eds) (1994) *At Risk: natural hazards, people's vulnerability and disasters*. Routledge, London.

Bloch, M. (1939) *Feudal Society* (two vols, translated from French by L.A. Manyou 1962), Vol. 1. Routledge and Kegan Paul, London.

Bodley, J.H. (2002) Anthropology and global environmental change. In Munn, T. (ed.) *Encyclopedia of Global Environmental Change* (five vols), Vol. 1. Wiley, Chichester, 1–5.

Bonnifield, M.P. (1979) *The Dust Bowl: men, dirt, and depression*. University of New Mexico Press, Albuquerque (NM).

Borroto, R.J. (1998) Global warming, rising sea-level, and growing risk of cholera incidence: of the literature and evidence. *GeoJournal* 44(2), 111–20.

Boserüp, E. (1965) *The Conditions of Agricultural Growth: the economics of agrarian change under population pressure*. Allen and Unwin, London.

Boserüp, E. (1990) *The Economic and Demographic Relationships in Development*. Johns Hopkins University Press, Baltimore (MD).

Botkin, D.B. (1990) *Discordant Harmonies: a new ecology for the twenty-first century*. Oxford University Press, Oxford.

Bradley, R.S, and P.D. Jones (eds) (1992) *Climate Since A.D. 1500* (revised and reprinted 1995). Routledge, London.

Brody, H. (2000) *The Other Side of Eden: hunter-gatherers, farmers and the shaping of the world*. Faber, London.

Broeker, W.S., Sutherland, S. and Peng T.-H. (1999) A possible 20th-century slowdown of Southern Ocean deepwater formation. *Science* 286, 1132–5.

Brooks, C.E.P. (1926) *Climate Through the Ages: a study of the climatic factors and their variation* (2nd edn 1942). Ernest Benn, London.

Brooks, C.E.P. (1950) *Climate in Everyday Life*. Ernest Benn, London.

Brown, L.R. (Project Director) (2001) *State of the World: a WorldWatch Institute report on progress towards a sustainable society*. W.W. Norton, New York (NY).

Bruce, T.P., Lee, H. and Haites, E.F. (eds) (1996) *Economic and Social Issues of Climate Change*. Cambridge University Press, Cambridge.

Bryant, E. (1997) *Climate Process and Change*. Cambridge University Press, Cambridge.

Bryant, E. (2002) *Tsunamis: the underrated hazard*. Cambridge University Press, Cambridge.

Bryson, R.A. and Murray, T.J. (1977) *Climates of Hunger: mankind and the world's changing weather*. University of Wisconsin Press, Madison (WS).

Budyko, M.I. (1974) *Climate and Life*. Academic Press, London.

Burroughs, W.J. (1997) *Does Weather Really Matter? The Social Implications of Climate Change*. Cambridge University Press, Cambridge.

Burton, I., Kates, R.W. and White, G.W. (1978) *The Environment as Hazard*. Oxford University Press, Oxford.

Calder, N. (1974) *The Weather Machine*. BBC Publications, London.

Cantor, N.F. (1997) *In the Wake of the Plague: the black death and the world it made*. Simon and Schuster, London.

Cartwright, F.F. and Biddis, M. (2001) *Diseases and History: the influence of disease in studying the great events of history*. Sutton, London.

Caviedes, C. (2001) *El Niño in History: storming through the ages*. University Press of Florida, Gainesville (FL).

Chagnon, S.A. (ed.) (2000) *El Niño 1997–1998: the climate event of the century*. Oxford University Press, Oxford.

Chambers, N., Simmons, C. and Wackernagel, M. (2000) *Sharing Nature's Interest: ecological footprints as an indicator of sustainability*. Earthscan, London.

Chappell, J.E. (1970) Climatic change reconsidered: another look at 'The Pulse of Asia'. *Geographical Review* 60, 347–73.

Charlesworth, J.K. (1957) *Quaternary Era* (two vols), Vol. 1. Arnold, London.

Chasek, P.S. (ed.) (2002) *The Global Environment in the Twenty-First Century: prospects for international cooperation*. United Nations University Press, Tokyo.

CIA (1974) *A Study of Climatological Research as it Pertains to Intelligence Problems* (Working Paper). Office of Research and Development, Central Intelligence Agency, Washington (DC).

Claiborne, R. (1970) *Climate, Man and History*. Angus and Robertson, London.

Clements, F.E. (1916) Plant succession. An analysis of the development of vegetation. *Carnegie Institute Publication* No. 242. Carnegie Institute, Washington (DC).

Coburn, K. (1964) *The Interpretation of Man and Nature*. Oxford University Press, Oxford.

Comfort, L., Wisner, B., Cutter, S., Pulwarty, R., Hewitt, K., Oliver-Smith, A., Wiener, J., Fordham, M., Peacock, W. and Krimgold, F. (1999) Reframing disaster policy: the global evolution of vulnerable communities. *Environmental Hazards* 1(1), 39–44.

Conte, F.P. (1995) The Aral Sea Basin. A man-made environmental catastrophe. *GeoJournal* 35(1), 33–4.

Conway, G. (2000) Food for all in the 21st century. *Environment* 42(1), 9–18.

Cooper, R. and Layard, R. (eds) (2001) *What the Future Holds: insights from social science*. MIT Press, Cambridge (MS).

Council on Environmental Quality and the Department of State (1982) *The Global 2000 Report to the President: entering the twenty-first century* (published in the USA in 1980) (two vols). Penguin, Harmondsworth.

Couper-Johnston, R. (2000) *El Niño: the weather phenomenon that changed the world*. Hodder and Stoughton, London.

Courtillot, V. (2000) *Evolutionary Catastrophes: the science of mass extinction.* Cambridge University Press, Cambridge.

Crane, A.T. (1990) Physical vulnerability of electrical systems to natural disaster and sabotage. *Terrorism* 13(3), 189.

Croll, E. and Parkin, D. (eds) (1992) *Bush Base: Forest Farm: culture, environment and development.* Routledge, London.

Crosby, A.W. (1972) *The Columbian Exchange, Biological and Cultural Consequences of 1492.* Greenwood Press, Westport (CN).

Crosby, A.W. (1986) *Ecological Imperialism: the biological expansion of Europe, 900–1900.* Cambridge University Press, Cambridge.

Crowley, T.J. and North, G.R. (1991) *Palaeoclimatology.* Oxford University Press, New York (NY).

Crumley, C.L. (ed.) (1994) *Historical Ecology: cultural knowledge and changing landscapes.* School of American Research Press, Santa Fe (NM).

Cullins, K.J. and Weiner, J.S. (1977) *Human Adaptability.* Taylor and Francis, London.

Currie, I. (1996) *Frosts, Freezes, and Fairs.* Frosted Earth, Coulsden (Surrey, UK).

Cutter, S.L. (1996) Vulnerability for environmental hazards. *Progress in Human Geography* 20(4), 529–39.

Cutter, S.L., Mitchell, T. and Scott, M.S. (2000) Revealing the vulnerability of people and places: a case study of Georgetown County, South Carolina. *Annals of the Association of American Geographers* 90(4), 713–37.

Dansgaard, W., Johnsen, S.J., Reeh, N., Gundestrup, N., Clausen, H.B. and Hammer, C.U. (1975) Climate changes, Norsemen, and modern man. *Nature* 255(5506), 271–87.

Davis, M. (2001) *Late Victorian Holocausts.* Verso, London.

Dawson, A.G. (1992) *Ice Age Earth: late Quaternary geology and climate.* Routledge, London.

de Haan, L.J. (2000) Globalization, localization and sustainable livelihood. *Sociologia Ruralis* 40(3), 339– 65.

de Vries, J. (1980) Measuring the impact of climate on history: the search for appropriate methodologies. *Journal of Interdisciplinary History* 10(4), 599–630.

Deighton, L. (2001) The Lost City. *New Scientist* 169(2273), 30–5.

Delano Smith, C. and Parry, M. (eds) (1981) *Consequences of Climatic Change.* Papers relating to the Historical Geography Research Group, Institute of British Geographers (held at the University of Nottingham, 11–15 January 1980). Nottingham University, Nottingham.

Diamond, J. (1997) *Guns, Germs and Steel: a short history of everybody for the last 13,000 years.* Jonathan Cape, London.

Dolby, D. and Harrison Church, R.J. (eds) (1973) *Drought in Africa.* Centre for African Studies, SOAS, University of London, London.

Dooge, J.C.I., Goodman, G.T., laRiviere, J.W.M., Marton- Lefévre, J., O'Riordan, T. and Praderie, F. (eds) (1992) *ASCEND21. An Agenda of*

Science for Environmental and Development into the 21st Century (ICSU). Cambridge University Press, Cambridge.

Döös, Bo. R. (1997) Can large-scale environmental migrations be predicted? *Global Environmental Change* 7(1), 41–61.

Döös, Bo R. and Shaw, R. (1999) Can we predict the future food production? A sensitivity analysis. *Global Environmental Change* 9(4), 261–83.

Dow, K. (1992) Exploring differences in our common future(s) – the meaning of vulnerability to global environmental change. *Geoforum* 23(3), 417.

Drury, S. (1999) *Stepping Stones: the making of our homeworld*. Oxford University Press, Oxford.

Dubos, R. (1965) *Man Adapting*. Yale University Press, New Haven (CN).

Dubos, R. (1973) *So Human an Animal*. Sphere Books (Abacus), London (published in USA, 1968).

Eden, S. (1998) Environmental issues: knowledge, uncertainty and the environment. *Progress in Human Geography* 22(3), 425–32.

Eisenberg, E. (1998) *The Ecology of Eden: humans, nature and human nature*. A.A. Knopf, New York (NY) and Macmillan (Picador), London.

El-Hinnawi, E. (1985) Environmental Refugees. UNEP, Unpublished Report. United Nations Environment Program, POB 30552, Nairobi, Kenya (40 pp.).

Ellen, R. (1982) *Environment, Subsistence and Systems: the ecology of small-scale social formations*. Cambridge University Press, Cambridge.

Emerson, R.W. (1954) *Five Essays on Man and Nature*. AHM Publishing Corp., Northbrook (IL).

Eyre, S.R. (1978) *The Real Wealth of Nations*. Edward Arnold, London.

Eyre, S.R. and Jones, G. (eds) (1966) *Geography as Human Ecology: methodology by example*. Arnold, London.

Fagan, B. (1999) *Floods, Famines, and Emperors: El Niño and the fate of civilizations*. Basic Books, New York (NY).

Fagan, B. (2000) *The Little Ice Age: how climate made history 1300–1850.* Basic Books, New York (NY).

Fagan, B.M. (1990) *Journey from Eden: the peopling of our world*. Thames and Hudson, London.

Fagan, B.M. (1995) *People of the Earth: an introduction to world prehistory* (eighth edition). HarperCollins, New York (NY).

Faust, B.B. (2001) Maya environmental successes and failures in the Yucatan Peninsula. *Environmental Science and Policy* 4(2), 153–69.

Febvre, L. (1924) *A Geographical Introduction to History*. Routledge and Kegan Paul, London.

Flenley, J.R. (1979) *The Equatorial Rainforest: a geological history*. Butterworths, London.

Flenley, J.R. (1998) Tropical forests under the climates of the last 30,000 years. *Climate Change* 39(2–3), 177–97.

Forde, C.D. (1934) *Habitat, Economy and Society: a geographic introduction to ethnology*. Methuen, London.

Fredskild, B. (1973) *Studies in the Vegetation History of Greenland: palaeobotanical investigations of some Holocene lake and bog deposits.* (Meddelelser om Gronland Bd 198, Nn 4). C.A. Reitzels Forlag, Copenhagen.

Frenkel, S. (1994) Old theories in new places? Environmental determinism and bioregionalism. *Professional Geographer* 46(1), 289–99.

Freyre, G. (1959) *New World in the Tropics: the culture of modern Brazil.* (Published in 1980, Westport, Greenwood Press.)

Friedrich, W.L. (2000) *Fire in the Sea: the Santorini Volcano: natural history and the legend of Atlantis* (translated into English from the German by A.R. McBirney). Cambridge University Press, Cambridge.

Fuchs, R.J., Brennan, E., Chamie, J., Lo, F.-C. and Uitto, J.I. (eds) (1994) *Mega-City Growth and the Future.* United Nations University Press, Tokyo.

Galbraith, J.K. (1951) Conditions for economic change in underdeveloped countries. *Journal of Farm Economics* 33(Nov. 1951).

Geertz, C. (1963) *Agricultural Involution: the process of ecological change in Indonesia.* University of California Press, Berkeley (CA).

Glacken, C.J. (1967) *Traces on the Rhodian Shore: nature and culture in western thought from ancient times to the end of the eighteenth century.* University of California Press, Berkeley (CA).

Glade, T., Albini, P. and Francés, F. (eds) (2001) *The Use of Historical Data in Natural Hazards Assessments.* Kluwer, Dordrecht

Glantz, M.H. (ed.) (1994) *Drought Follows the Plough: cultivating marginal areas.* Cambridge University Press, Cambridge.

Glantz, M.H. (ed.) (1999) *Creeping Environmental Problems and Sustainable Development in the Aral Sea Basin.* Cambridge University Press, Cambridge.

Glantz, M.H., Rubenstein, A. and Zonn, I. (1993) Tragedy in the Aral Sea basin: looking back to plan ahead? *Global Environmental Change* 3(3), 174–98.

Glantz, M.H. (2002) *La Niña and its impacts: facts and speculation.* United Nations University Press, Tokyo.

Gottfried, R.S. (1983) *The Black Death: natural and human disaster in medieval Europe.* R. Hale, London.

Goudie, A.S. (1977) *Environmental Change.* Clarendon Press, Oxford.

Goudie, A.S. (ed.) (2002) *Encyclopedia of Global Change: environmental change and human society* (two vols). Oxford University Press, Oxford.

Gould, S.J. (1989) *Wonderful Life: the Burgess Shale and the nature of history.* Hutchinson Radius, London.

Gould, S.J. (2000) *The Lying Stones of Marrakech: penultimate reflections in natural history.* Jonathan Cape, London.

Grattan, J.P. and Pyatt, F.B. (1999) Volcanic eruptions, dry fogs, and the European palaeoenvironmental record: localised phenomena or hemispheric impacts? *Global and Planetary Change* 21(1–3), 173–9.

Gribbin, J. (1976a) *Forecasts, Famines and Freezes: climate and man's future.* Wildwood House, London.

Gribbin, J. (1976b) *Our Changing Universe.* Methuen, London.

Gribbin, J. (1978) *The Climatic Threat: what's wrong with our weather?* Fontana, London.

Gribbin, J. and Gribbin, M. (1998) Bolts from the blue ('Inside Science' supplement). *New Scientist* **159**(2151), 4 pp.

Grove, A.T. (1996) The historical context: before 1850. In C.J. Brandt and J.B. Thornes (eds) *Mediterranean Desertification and Land Use.* Wiley, Chichester, 13–28.

Grove, J.M. (1988) *The Little Ice Age.* Cambridge University Press, Cambridge.

Grove, R.H. and Chappell, J. (eds) (2000) *El Niño: History and Crisis: studies from the Asia-Pacific region.* White Horse Press, Cambridge.

Haberle, S.G. and Chepstow Lusty, A. (2000) Can climate change influence cultural development? A view through time. *Environment and History* **6**(3), 349–69.

Hadfield, P. (1991) *The Coming Tokyo Earthquake: sixty seconds that will change the world.* Charles E. Tuttle, Boston (MA).

Hamaker, J.D. (1982) *The Survival of Civilization* (contributions by D.A. Weaver). Hamaker-Weaver Publications, Box 457 Potterville 48876 (MI).

Hancock, G. (2002) *Underworld: flooded kingdoms of the ice age.* Michael Joseph (Penguin Books), London.

Hantoro, W.S., Faure, H., Djuwansah, R., Faure-Denard, L. and Pirazzoli, P.A. (1995) The Sunda and Sahul continental platform: lost land of the last glacial continent in S.E. Asia. *Quaternary International* **29–30**, 129–34.

Harding, J. (2000) *The Uninvited: refugees at the rich man's gate.* Profile and the London Review of Books, London.

Hardoy, J., Mitlin, D. and Satterthwaite, D. (2002) *Environmental Problems in an Urbanizing World: finding solutions for cities in Africa, Asia and Latin America.* Earthscan, London.

Harris, M. (1965) The myth of the sacred cow. In A. Leeds and A.P. Vayda (eds) *Man, Culture and Animals: the role of animals on human adjustments.* American Association for the Advancement of Science Publication No. 78. AAAS, Washington (DC).

Harris, M. (1971) *Culture, Man and Nature: an introduction to general anthropology.* Crowell, New York (NY).

Harris, M. (1974) *Cows, Pigs, Wars and Witches: riddles of culture.* Random House, New York (NY).

Harris, M. (1992) The cultural ecology of India's sacred cattle. *Current Anthropology* **7**(1), 51–66.

Harrison, P. (1979) The curse of the tropics. *New Scientist* **84**(1182), 602–4.

Harvey, L.D.D. (2000a) *Global Warming: the hard science*. Prentice-Hall, New York (NY).

Harvey, L.D.D. (2000b) *Climate and Global Environmental Change*. Prentice-Hall, New York (NY).

Hewitt, K. and Hare, F.K. (1973) *Man and the Environment: conceptual frameworks*. Association of American Geographers, Washington (DC).

Hills, J.G. and Goda, M.P. (1999) Damage from comet-asteroid impacts with Earth. *Physica D: Nonlinear Phenomena* 133(1–4), 189–98.

Holdridge, L.R. (1967) *Life Zone Ecology* (revised edn). Tropical Science Center, San José (Costa Rica).

Holling, C.S. (1978) *Adaptive Environmental Assessment and Management*. Wiley, Chichester.

Honigsbaum, M. (2001) *The Fever Trail: the hunt for the cure for malaria*. Macmillan, London.

Hoyle, F. (1981a) The next ice age. *The Sunday Times Magazine*, 7 June 1981, 18–26.

Hoyle, F. (1981b) *Ice*. Hutchinson, London.

Huntington, E. (1915) *Civilization and Climate*. Yale University Press, New Haven (CN).

Huntington, E. (1919) *The Pulse of Asia: a journey in central Asia illustrating the geographical basis of history* (2nd edn; 1st edn 1912). Constable, London.

Huntington, E. (1945) *Mainsprings of Civilization*. Wiley, New York (NY).

Imbrie, J. and Imbrie, K.P. (1979) *Ice Ages: solving the mystery*. Macmillan, Basingstoke (UK) and Enslow Publishers, Short Hills (NJ).

Ingram, M.J., Farmer, G. and Wigley, M.L. (eds) (1981) *Climate and History: studies in past climates and their impact on man*. Cambridge University Press, Cambridge.

International Federation of Red Cross and Red Crescent Societies (2001) *World Disasters Report 2000*. Published by the Red Cross, Geneva (distributed by Europspan, ISBN 92 9139 066 6, available from wdorders@ifrc.org).

Irwin, A. (2001) *Sociology and the Environment: a critical introduction to society, nature and knowledge*. Polity Press, Cambridge.

James, P. (1995) *The Sunken Kingdom: the Atlantis mystery solved*. Jonathan Cape, London.

John, B.S. (ed.) (1979) *The Winters of the World: earth climate under the ice ages*. David and Charles, Newton Abbot (UK).

Johnson, D.L. and Lewis, L.A. (1995) *Land Degradation: creation and destruction*. Blackwell, Oxford.

Jones, G. (1964) *The Norse Atlantic Saga: being the Norse voyages of discovery and settlement to Iceland, Greenland, America*. Oxford University Press, London.

Jones, N. (2001a) Never say die: as the world warms, what's in store for the Amazon rainforest? *New Scientist* 169(2275), 36–9.

Jones, N. (2001b) The monster in the lake. *New Scientist* **169**(2283), 36–40.

Jones, T.L., Brown, G.M., Raab, L.M., McVickar, J.L., Spalding, W.G., Kennett, D.J., York, A. and Walker, P.L. (1999) Environmental imperatives reconsidered. *Current Anthropology* **40**(2), 137–70.

Kahn H., Brown, W. and Martel, L. (1976) *The Next 200 Years*. Based on the 'Hudson Report' from the Hudson Institute (New York). Published in the UK, 1978, by Sphere Books, London.

Kaser, G. (1999) A review of the modern fluctuations of tropical glaciers. *Global and Planetary Change* **22**(1–4), 93–103.

Kasperson, J.X. and Kasperson, R.E. (eds) (2001) *Global Environmental Risk*. Earthscan, London.

Kasperson, J.X., Kasperson, R.E. and Turner, B.L. (1996) Regions at risk: exploring environmental criticality. *Environment* **38**(10), 4–15, 26–9.

Kates, R.W. and Clark, W.C. (1996) Environmental surprise: expecting the unexpected. *Environment* **38**(1), 6–11, 28–34.

Kellert, S. and Wilson, E.O. (eds) (1993) *The Biophilia Hypothesis*. Island Press, Washington (DC).

Kennedy, P. (1993) *Preparing for the Twenty-First Century*. HarperCollins, London.

Keys, D. (1999) *Catastrophe: an investigation into the origins of the modern world*. Century Books, London.

Kiel, R., Bell, V.J., Penz, P. and Fawcette, L. (eds) (1998) *Political Ecology: global and local*. Routledge, London.

Kondratyev, K.Ya. (1988) *Climate Shocks: natural and anthropogenic*. Wiley, New York (NY).

Kottak, C.P. (1989) The new ecological anthropology. *American Anthropologist* **101**(1), 23–35.

Krech, S. III (2001) *The Ecological Indian: myth and history*. Norton, New York (NY).

Kunreuther, H. and Roth, Sr.R. (1998) *Paying the price: the status and rate of insurance against natural disasters in the United States*. Joseph Henry Press, Washington (DC).

Lamb, H.H. (1977) *Climate History and the Future*. Methuen, London.

Lamb, H.H. (1982) *Climate, History and the Modern World* (2nd edn published 1995). Methuen London.

Lamb, H.H. (1988) *Weather, Climate and Human Affairs*. Routledge, London.

Landes, D. (1998) *The Wealth and Poverty of Some Nations: why some are so rich and some so poor*. W.W. Norton and Co., Boston (MS) and Little, Brown and Co., London.

Lash, S., Szerszynski, B. and Wynn, B. (eds) (1996) *Environment and Modernity: towards a new ecology*. Sage, London.

Le Roy Ladurie, E. (1972) *Times of Feast, Times of Famine: a history of climate since the year 1000* (translated from the French by B. Brag).

George Allen and Unwin, London (1971 edn, Doubleday, New York, NY; originally published in French, 1967, *Histoire du climat dupuis l'an mil*. Flammarion, Paris).

Leach, M. and Mearns, E. (eds) (1996) *The Lie of the Land: challenging received wisdom on the African environment* (published for the International African Institute). James Currey, London and Indiana University Press, Bloomington (IN).

Lee, D.H.K. (1957) *Climate and Economic Development in the Tropics*. Harper Brothers, New York (NY).

Lewis, J. (1999) *Development in Disaster-Prone Places: studies in vulnerability*. Intermediate Technology Publications, London.

Lewis, M.W. (1992) *Green Delusions: an environmentalist critique of radical environmentalism*. Duke University Press, Durham (NC).

Ligon, L. (2001) Smallpox: its history and reemergence as a weapon of biological warfare. *Seminars in Pediatric Infectious Diseases* 12(1), 71–80.

Little, P.E. (1983) Environments and environmentalisms in anthropological research. *Annual Review of Anthropology* 28, 253–84.

Lomborg, B. (2001) *The Sceptical Environmentalist: measuring the real state of the world* (1st edn published in Danish 1998). Cambridge University Press, Cambridge.

Longshore, D. (2000) *Encyclopedia of Hurricanes, Typhoons and Cyclones*. Fitzroy, Dearborn (MS).

Lorenzoni, I., Jordan, A., Hulme, M., Kerry Turner, R. and O'Riordan, T. (2000) A co-evolutionary approach to climate change impact assessment: Part I. Integrating socio-economic and climate change scenarios. *Global Environmental Change* 10(1), 57–68.

Lovelock, J.E. (1979) *Gaia: a new look at life on earth*. Oxford University Press, Oxford.

Lovelock, J.E. (1988) *The Ages of Gaia: a bibliography of our living earth*. W.W. Norton, London (2nd edn 1995, Oxford University Press, Oxford).

Lovelock, J.E. (1992) *Gaia: the practical science of planetary medicine*. Gaia Books, London.

Lutz, W. (ed.) (1996) *The Future Population of the World: what can we assume today?* (revised edn; 1st edn 1994). Earthscan, London.

Maarleveld, G.C. and Van der Hammen, Th. (1959) The correlation between upper Pleistocene pluvial and glacial stages. *Geologie en Mijnbouw* (new series) 21e, 40–5.

McCrone, J. (2000) Fired up. *New Scientist* 166(2239), 30–3.

McGregor, J. (1994) Climate change and involuntary migration: implications for food security. *Food Policy* 19(2), 120–32.

McGuire, B. (1999) *Apocalypse: a natural history of global disasters*. Cassell and Co., London.

McGuire, W.J., Griffiths, D.R., Handcock, P.L. and Stewart, I.S. (2000) *The Archaeology of Geologic Catastrophes* (Geological Society Special Publication No. 171). Geological Society, London.

Mackenzie, D. (2001) Ring a ring o'roses, a pocket full of posies, atishoo! atishoo! We all fall down. *New Scientist* 172(2318), 34–8.

McKibbin, B. (1990) *The End of Nature.* Viking Penguin, New York (NY).

McMichael, A.J., Haines, A., Sloof, R. and Kovats, S. (eds) (1996) *Climate Change and Human Health* (An Assessment by a Task Group on Behalf of the WHO, WMO and the UNEP). World Health Organization, Geneva.

McMichael, T. (2001) *Human Frontiers, Environments and Disease.* Cambridge University Press, Cambridge.

McNeill, J.R. (2000) *Something New Under the Sun: an environmental history of the twentieth century.* Penguin, London.

Magnusson, M. and Palsson, H. (1965) *The Vinland Sagas: the Norse discovery of America.* Penguin Books, Harmondsworth.

Malthus, T.R. (1798) *First Essay on Population.* Penguin, Harmondsworth.

Mannion, A.M. (1994) The new environmental determinism. *Environmental Conservation* 21(3), 7–9.

Mannion, A.M. (1999) *Natural Environmental Change: the last three million years.* Routledge, London.

Markham, S.F. (1942) *Climate and the Energy of the Nations* (2nd edn 1947). Oxford University Press, Oxford.

Marshall, T. (2001) The drowning wave. *New Scientist* 168(2259), 26–30.

Marshall, T. (2002) There she blows. *New Scientist* 173(2325), 29–31.

Martin, G. (1973) *Ellsworth Huntington. His Life and Thought.* Archon Books, Hamden (CT).

Martin, P.S. (1984) Prehistoric overkill: the global model. In P.S. Martin and R.G. Klein (eds) *Quaternary Extinctions: a prehistoric revolution.* University of Arizona Press, Tucson (AZ), 553–73.

Martin, P.S. and Wright, H.E.W. (eds) (1967) *Pleistocene Extinctions: the search for a cause.* Yale University Press, New Haven (CN).

Mass, C.F. and Portman, D.A. (1989) Major volcanic eruptions and climate: a critical evaluation. *Journal of Climate* 2, 566–93.

Mastin, L.G. and Witter, J.B. (2000) The hazards of eruptions through lakes and seawater. *Journal of Vulcanology and Geothermal Research* 97(1–4), 195–214.

Matthews, J.A. (ed.) (2001) *The Encyclopaedic Dictionary of Environmental Change.* Arnold, London.

Meadows, D.H., Meadows, D.L., Randers, J. and Behrens, W.W. III (1972) *The Limits to Growth: a report from the Club of Rome's Project on the Predicament of Mankind.* Universe Books, New York (NY).

Meadows, D.H., Meadows, D.L. and Randers, J. (1992) *Beyond the Limits: global collapse or a sustainable future?* Earthscan, London.

Merchant, C. (1989) *The Death of Nature.* Harper and Row, San Francisco (CA).

Metzger, P., D'Ercole, R. and Sierra, A. (1999) Political and scientific uncertainties in volcanic risk management: the yellow alert in Quito in October 1998. *GeoJournal* **49**(2), 213–22.

Mickin, P.P. (2001) Desiccation of the Aral Sea: a water management disaster in the Soviet Union. *Science* **241**(4870), 1170–6.

Miller, M. (2001) Mammoth mystery. *New Scientist* **170**(2289), 32–5.

Milliman, J.D., Broadus, J.M. and Gable, F. (1989) Environmental and economic implications of rising sea level and subsiding deltas: the Nile and Bengal examples. *Ambio* **XVIII**(6), 340–5.

Millington, A.C. and Pye, K. (eds) (1994) *Environmental Change in Drylands: biogeographical and geomorphological perspectives.* Wiley, Chichester.

Mitchell, J.K. (1995) Coping with natural hazards and disasters in megacities: perspectives on the twenty-first century. *GeoJournal* **37**(3), 303–11.

Mitchell, J.K. (1999) Megacities and natural disasters: a comparative analysis. *GeoJournal* **49**(2), 137–42.

Mitchell, J.K. (ed.) (2001) *Crucibles of Hazard: mega-cities and disasters in transition.* United Nations University Press, Tokyo.

Mooley, D.A. and Pant, G.B. (1981) Droughts in India over the last 200 years: their socio-economics and remedial measures for them. In T.M.L. Wigley *et al.* (eds) *Climate and History.* Cambridge University Press, Cambridge, 465–78.

Moran, E. (1984) *The Ecosystem Concept in Anthropology.* Westview Press, Boulder (CO).

Moran, E. (1990) *The Ecosystem Approach in Anthropology.* University of Michigan Press, Ann Arbor (MI).

More, T. (1516) *Utopia* (English translation published 1965). Penguin, Harmondsworth.

Morgan, E. (1972) *The Descent of Women.* Stein and Day, New York (NY).

Mortimore, M. (1989) *Adapting to Drought: farmers and desertification in West Africa.* Cambridge University Press, Cambridge.

Moss, N. (2001) *Managing the Planet: the politics of the new millennium.* Earthscan, London.

Muir, H. (2001) Target Earth. *New Scientist* **169**(2280), 40–4.

Munn, T. (ed.) (2002) *Encyclopedia of Global Change* (five vols). Wiley, Chichester.

Myers, N. (1993) Environmental refugees in a globally warmed world. *Bioscience* **43**(11), 752–61.

Myers, N. (1995) Environmental unknowns. *Science* **269**(5222), 358–60.

Myers, N. and Kent, J. (1995) *Environmental Exodus: an emergent crisis in the global arena.* Climate Institute, Washington (DC).

Napier, W.M. and Clure, V.M. (1979) A theory of terrestrial catastrophism. *Nature* **282**(5738), 455–9.

Netting, R. McM. (1977) *Cultural Ecology.* Cummings Publishing Co., Reading (MS).

Nisbet, E.G. (1991) *Leaving Eden: to protect and manage the Earth.* Cambridge University Press, Cambridge.

Norgaard, R.B. (1984) *Development Betrayed: the end of progress and a coevolutionary revisioning of the future.* Routledge, London.

O'Hanlon, L. (2001) Quakin' all over. *New Scientist* 170(2258), 30–3.

Officer, C. and Page, J. (1998) *Tales of the Earth: paroxysms and perturbations of the blue planet.* Oxford University Press, New York (NY).

Ogilvie, A.E.J. and Jónsson, T. (eds) (2001) *The Iceberg in the Mist: northern research in pursuit of a Little Ice Age.* Kluwer, Dordrecht.

Oldfield, F. (2000) Out of Africa. *Nature* 403(6788), 370–1.

Oppenheimer, S. (1999) *Eden in the East: the drowned continent of South East Asia.* Phoenix Books, London.

Pain, S. (1994) 'Rigid' cultures caught out by climate change. *New Scientist* 141(1915), 13.

Paldam, M. (2000) Social capital: one or many? Definition and measurement. *Journal of Economic Surveys* 14(5), 629–53.

Park, R.E. and Burgess, E.W. (1924) *Introduction to the Science of Sociology.* University of Chicago Press, Chicago (IL).

Parker, D. and Mitchell, J.K. (1995) Disaster vulnerability of megacities: an expanding problem that requires rethinking and innovative responses. *GeoJournal* 37(3), 295–301.

Parry, M. and Carter, T. (1998) *Climate Impact and Adaptation Assessment: a guide to the IPCC Approach.* Earthscan, London.

Parry, M.L. (1978) *Climatic Change, Agriculture and Settlement.* Dawson, Folkestone.

Parry, M.L., Rosenzweig, C., Iglesias, A., Fischer, G. and Livermore, M. (1999) Climate change and world food security: a new assessment. *Global Environmental Change* 9(9), S51–61.

Pearce, F. (2000a) Feel the pulse. *New Scientist* 2254(167), 30–3.

Pearce, F. (2000b) Unheeded Warnings: predictions of floods in Mozambique were ignored. *New Scientist* 165(2229), 21.

Pearce, F. (2001) Coat of many colours. *New Scientist* 172(2316), 36–9.

Pearce, F. (2002) The rainmaker. *New Scientist* 173(2317), 41–3.

Pearson, R. (1978) *Climate and Evolution.* Academic Press, New York (NY).

Pepper, D. (1984) *The Roots of Modern Environmentalism.* Croom Helm, London.

Pfister, C., Brázdil, R., Glaser, R., Barriendos, M., Camuffo, D., Deutsch, M., Dobrovolni, P., Enzi, S., Guidoboni, E., Kotyza, O., Militzer, S., Rácz, L. and Rodrigo, F.S. (1999) Documentary evidence on climate in sixteenth-century central Europe. *Climatic Change* 43(1), 55–110.

Pitts, M. and Roberts, M. (1998) *Fairweather Eden: life in Britain half a million years ago as revealed by the excavations at Boxgrove.* Arrow Books (Random House), London.

Ponting, C. (1991) *A Green History of the World.* Sinclair-Stevenson, London.

Pratt, V., Howarth, J. and Brady, E. (2000) *Environmental Philosophy*. Routledge, London.

Pretty, J. and Ward, H. (2001) Social capital and the environment. *World Development* **29**(2), 209–27.

Pringle, H. (1997) Death in Norse Greenland. *Science* **275**(5302), 924–6.

Rabinowitz, D., Helin. E., Lawrence, K. and Pravdo, S. (2000) A reduced estimate of the number of kilometre-sized near-Earth asteroids. *Nature* **403**(6766), 165–6.

Ramlogam, R. (1996) Environmental refugees. *Environmental Conservation* **23**(1), 81–8.

Ratzel, F. (1889) *The History of Mankind* (first published in German in 1882; English translation from the German by A.J. Butler). Macmillan, London.

Raudzens, G. (ed.) (2001) *Technology, Disease and Colonial Conquests, Sixteenth to Eighteenth Centuries: essays reappraising the guns and germs theory*. Brill Academic Publishers, Leiden (and Herndon, VA).

Raup, D. (1986) *The Nemesis Affair*. Norton, New York (NY).

Rayner, S. and Malone, E.L. (eds) (1998) *Human Choice and Climate Change* (four vols), Vol. 1: the societal framework. Battelle Press, Columbus (OH).

Reiter, P. (2000a) Opinion interview: biting back. *New Scientist* **167**(2257), 41–3.

Reiter, P. (2000b) From Shakespeare to Defoe: malaria in England in the Little Ice Age. *Emerging Infectious Diseases* (National Center for Disease Control and Prevention, USA) **6**(1), 14 pp. Also published in *Emerging Infectious Diseases* **6**(1), 1–11.

Rivers, P. (1988) *The Stolen Future: how to rescue the Earth for our children*. Green Print, Basingstoke (UK).

Rosen, P. (2000) Coping with bioterrorism. *British Medical Journal* **320**(7227), 71–2.

Ruddiman, W.F. (2001) *Earth's Climate: past and future*. W.H. Freeman, New York (NY).

Ryan, W. and Pitman, W. (2000) *Noah's Flood*. Simon and Schuster, New York (NY).

Sadler, J.P. and Grattan, J.P. (1999) Volcanoes as agents of past environmental change. *Global and Planetary Change* **21**(1–3), 181–96.

Sale, K. (1985) *Dwellers in the Land: the bioregional vision*. Sierra Club, 730 Polk Street, San Francisco (CA). Re-published in 1991 by New Society Publishers, Santa Cruz (CA).

Sale, K. (1991) *Dwellers in the Land: the bioregional vision*. New Society Publishers, Philadelphia (PA). First published 1985 by University of Georgia Press, Philadelphia (PA).

Sauer, C.O. (1952) *Agricultural Origins and Dispersals*. American Geographical Society, New York (NY).

Sauer, C.O. (1963) *Land and Life: a selection from the writings of Carl*

Ortwin Sauer (ed. J. Leighly). University of California Press, Berkeley (CA).

Schumacher, E.F. (1973) *Small is Beautiful: a study of economics as if people really mattered.* Bland and Briggs, London.

Scoones, I. (ed.) (2001) *Dynamics & Diversity: soil fertility and farming livelihoods in Africa.* Earthscan, London.

Scott, A.C., Lomax, B.H., Collinson, M.E., Upchurch, G.R. and Beerling, D.J. (2000) Fire across the K-T boundary: initial results from the Sugarite Coal, New Mexico, USA. *Palaeogeography, Palaeoclimatology, Palaeoecology* 164(1–4), 151–65.

Scott, S. and Duncan, C.J. (2001) *Biology of Plagues: evidence from historical populations.* Cambridge University Press, Cambridge.

Semple, E.C. (1911) *Influences of Geographic Environment the Basis of Ratzel's System of Anthropo-Geography.* Henry Holt, New York (NY).

Shah, H.C. (1995) The increasing nature of global earthquake risk. *Global Environmental Change* 5(1), 65–7.

Shoemaker, E.M. (1998) Impact cratering through geologic time. *Journal of the Royal Astronomical Society of Canada* 92(6), 297–309.

Simmons, I.G. (1997) *Humanity and Environment: a cultural ecology.* Addison Wesley Longman, Harlow.

Sjoberg, L. (ed.) (1987) *Risk and Society: studies of risk generation and reactions to risk.* Allen and Unwin, London.

Skinner, B.F. (1972) *Beyond Freedom and Dignity and About Behaviouralism.* Jonathan Cape, London.

Slack, P. (ed.) (1999) *Environments and Historical Change* (The Linacre Lectures, 1998). Oxford University Press, Oxford.

Smith, K. (1992) *Environmental Hazards: assessing the risk and reducing disaster.* Routledge, London.

Spray, J.G., Kelley, S.P. and Rowley, D.B. (1998) Evidence for a late Triassic multiple impact event on Earth. *Nature* 392(6672), 171–3.

Steel, D. (1995) *Rogue Asteroids and Doomsday Comets.* Wiley, New York (NY).

Steward, J. (1955) *The Theory of Culture Change: the methodology of multilinear evolution.* University of Illinois Press, Urbana (IL).

Stone, C.D. (1993) *The Gnat is Older than the Man: global environmental and human agenda.* Princeton University Press, Princeton (NJ).

Stott, P. and Sullivan, S. (eds) (2000) *Political Ecology: science, myth and power.* Arnold, London.

Streets, D.G. and Glantz, M.H. (2000) Exploring the concept of climate surprise. *Global Environmental Change* 10(2), 97–107.

Takahashi, S. (1998) Social geography and disaster vulnerability in Tokyo. *Applied Geography* 18(1), 17–24.

Tate, J. (2000) Avoiding collisions: the Spaceguard Foundation. *Space Policy* 16(4), 261–5.

Thomas, K. (1983) *Man and the Natural World: changing attitudes in England 1500–1800*. Allen Lane, London (Penguin edn, 1984).

Tiffen, M. (1993) Productivity and environmental conservation under a rapid population growth: a case- study of Machacos District, Kenya. *Journal of International Development* 5(2), 207–24.

Toon, O.B., Pollack, J.B, Ackerman, T.P, Turco, R.P, McKay, C.P and Liu, M.S. (1982) Evolution of an impact-generated dust cloud and its effects on the atmosphere. *Special Papers of the Geological Society of America* No. 190, 187–200.

Toynbee, A. (1976) *Mankind and Mother Earth: a narrative history of the world*. Oxford University Press, Oxford.

Toynbee, A.J. (1934) *A Study of History* (three vols). (The 1946 abbreviated version was edited by D.C. Somervell.) Oxford University Press, London.

Tuchman, B.W. (1978) *A Distant Mirror: the calamitous 14th century*. Knopf, New York (NY).

Turekian, K.K. (1996) *Global Environmental Change: past, present, and future*. Prentice-Hall, Upper Saddle River (NJ).

Turner, F.J. (1962) *Frontier in American History* (first published 1920, Henry Holt and Co., New York (NY).

Uitto, J.I. (1998) The geography of disaster vulnerability in megacities – a theoretical framework. *Applied Geography* 18(1), 7–16.

Ulijaszek, S.J. and Huss-Ashmore, R.A. (eds) (1997) *Human Adaptability: past, present, and future*. Oxford University Press, Oxford.

Van Apeldoorn, G. Jan. (1981) *Perspectives on Drought and famine in Nigeria*. George Allen and Unwin, London.

Vandderbloemen, J.G. (2000) Are we determined to be possibilists or is it possible to be a determinist? Paper published on the Internet at http://delta.marine.ust.edu/~joevan/dtermin.html, accessed 25 May 2000.

van Ginkel, H., Barrett, B., Court, J. and Velasquez, J. (eds) (2001) *Human Development and the Environment: challenges for the United Nations in the new millennium*. United Nations University Press, Tokyo.

Vayda, A.F. (ed.) (1969) *Environment and Cultural Behavior: ecological studies in cultural anthropology*. Natural History Press, Garden City (NY).

Velakovsky, I. (1950) *Worlds in Collision*. Victor Gollancz, London.

Velakovsky, I. (1952) *Ages in Chaos*. Victor Gollancz, London.

Velakovsky, I. (1955) *Earth in Upheaval*. Victor Gollancz, London.

Vidal de la Blanche, P. (1918) *Principles of Human Geography*. Posthumously edited by E. de Martonne; English translation by M.T. Bingham. Constable Publishers, London (1926).

Voortman, R.L (1998) Recent historical climate change and its effect on land use in the eastern part of West Africa. *Physics and Chemistry of the Earth* 23(4), 385–91.

Walker, R., Landis, W. and Brown, P. (2001) Developing a region al ecological risk assessment: a case study of a Tasmanian agricultural catchment. *Human and Ecological Risk Assessment* 7(2), 417–39.

Walters, C.J. (1986) *Adaptive Management of renewable resources.* Macmillan, New York (NY).

Waple, A.M. (1999) The Sun–climate relationship in recent centuries: a review. *Progress in Physical Geography* 23(3), 309–28.

Ward, B. and Dubois, R. (1972) *Only One Earth: the care and maintenance of a small planet.* Penguin, Harmondsworth.

Watts, M.J. and Bohle, H.G. (1993) The space of vulnerability: the causal structure of hunger and famine. *Progress in Human Geography* 17(1), 43–67.

Watts, S. (1997) *Epidemics and History: disease, power, and imperialism.* Yale University Press, New Haven (CN).

Westing, A.H. (1992) Environmental refugees: a growing category of displaced persons. *Environmental Conservation* 19(3), 201–7.

Whitaker, D.W., Wasimi, S.A. and Islam, S. (2001) The El Niño-Southern Oscillation and long-range forecasting of flows in the Ganges. *International Journal of Climatology* 21(1), 77–87.

White, G.E. (1974) *Natural Hazards: local, national, global.* Oxford University Press, Oxford.

White, L. Jr (1967a) The historical roots of our ecological crisis. *Science* 155(3767), 1203–7.

White, L. Jr (1967b) The historical roots of our ecological crisis. In J. Barr (ed.) *The Environmental Handbook.* Ballantine, London.

Whittow, J. (1980) *Disasters: the anatomy of environmental hazards.* Pelican, Harmondsworth.

Wickramasinghe, C. (2001) *Cosmic Dragons: life and death on our planet.* Souvenir Press, London.

Wigley, T.M.L., Ingram, M.J. and Farmer, G. (eds) (1981) *Climate and History: studies in past climates and their impact on man.* Cambridge University Press, Cambridge.

Wijkman, A. and Timberlake, L. (1984) *Natural Disasters: acts of God or acts of man?* Earthscan, London.

Williams, W.M. and Ceci, S.J. (eds) (1999) *The Nature-Nurture Debate: the essential readings.* Blackwell, Malden (MA).

Wilson, E.O. (1975) *Sociobiology: the new synthesis.* Bellknap Press (Harvard University Press), Cambridge (MS).

Wilson, E.O. (1984) *Biophilia.* Harvard University Press, Cambridge (MS).

Woolfson, D. (2000) Humanity's evolutionary future. *World Futures* 55(5), 182–200.

World Bank (1992) *The World Development Report 1992: development and the environment.* Oxford University Press, Oxford.

World Commission on Environment and Development (1987) *Our Common Future* (the 'Brundtland Report'). Oxford University Press, Oxford.

Worster, D. (1979) *Dust Bowl: the Southern Plains in the 1930s.* Oxford University Press, New York (NY).

Worster, D. (ed.) (1988) *The Ends of the Earth: perspectives on modern environmental history.* Cambridge University Press, Cambridge.

Zangger, E. (2001) *The Future of the Past: archaeology in the twenty-first century.* Weidenfeld and Nicolson, London.

Zebrowski, E. Jr (2001) *The Last Days of St. Pierre: the volcanic disaster that claimed 30,000 lives.* Rutgers University Press, Piscataway (NJ).

Ziegler, P. (1990) *Black Death.* Penguin, Harmondsworth.

Index

Milton Keynes UK
Ingram Content Group UK Ltd.
UKHW040107071024
449327UK00019B/869

9 780340 764046